T5-BBP-691

GLACIERS, SEA ICE, AND ICE FORMATION

DYNAMIC EARTH

GLACIERS, SEA ICE, AND ICE FORMATION

EDITED BY JOHN P. RAFFERTY, ASSOCIATE EDITOR, EARTH SCIENCES

Britannica®
Educational Publishing

IN ASSOCIATION WITH

ROSEN
EDUCATIONAL SERVICES

Published in 2011 by Britannica Educational Publishing
(a trademark of Encyclopædia Britannica, Inc.)
in association with Rosen Educational Services, LLC
29 East 21st Street, New York, NY 10010.

First Edition

Britannica Educational Publishing
Michael I. Levy: Executive Editor
J.E. Luebering: Senior Manager
Marilyn L. Barton: Senior Coordinator, Production Control
Steven Bosco: Director, Editorial Technologies
Lisa S. Braucher: Senior Producer and Data Editor
Yvette Charboneau: Senior Copy Editor
Kathy Nakamura: Manager, Media Acquisition
John P. Rafferty: Associate Editor, Earth Sciences

Rosen Educational Services
Jeanne Nagle: Editor
Nelson Sá: Art Director
Cindy Reiman: Photography Manager
Matthew Cauli: Designer, Cover Design
Introduction by Theresa Shea

Library of Congress Cataloging-in-Publication Data

Glaciers, sea ice, and ice formation / edited by John P. Rafferty.—1st ed.
    p. cm.—(Dynamic earth)
"In association with Britannica Educational Publishing, Rosen Educational Services."
Includes bibliographical references and index.
ISBN 978-1-61530-119-5 (library binding)
1. Glaciers. 2. Sea ice. 3. Ice. I. Rafferty, John P. II. Series: Dynamic earth.
GB2403.2.G54 2010
551.31—dc22

                                                                        2010000226

Manufactured in the United States of America

CONTENTS

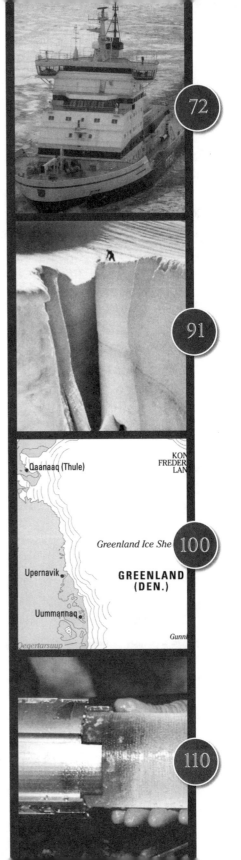

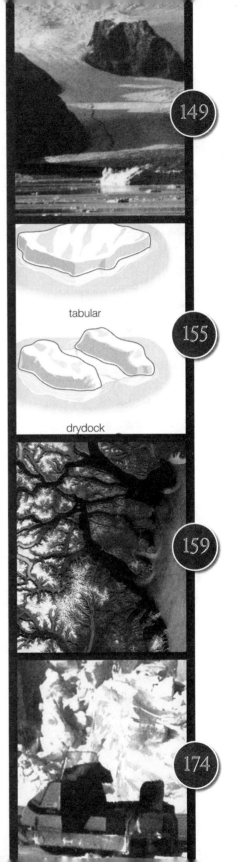

tabular

drydock

178

185

194

209

INTRODUCTION

Water is at once simple and complex. Its study begins with a single molecule. A water molecule consists of two hydrogen atoms and one oxygen atom, as noted by its chemical formula H_2O. In water's gaseous state, thermal energy—heat—enables single water molecules to float freely, mostly independently of one another. In water's liquid state, hydrogen atoms constantly form and break bonds with other hydrogen atoms, as exhibited by water's characteristic fluidity. In the solid phase, water's molecular bonds are dictated by the oxygen atoms, which form crystalline shapes, or ice.

On Earth, ice appears in many forms, from the ground-imbedded nature of permafrost to the towering majesty of mountainous glaciers. Each of these varied states is examined thoroughly in *Glaciers, Sea Ice, and Ice Formation*. Readers will virtually traverse pingos, rappel down cirques, and explore the frozen tundra of the Arctic and Antarctic as the complexities of this deceptively simple entity are revealed.

Unlike most materials, water's solid state is less dense than its liquid state, a factor that explains why ice floats above liquid water. People witness natural ice formations seasonally on lakes, ponds, and other small bodies of water. As average daily temperatures fall, thermal energy is released from liquid water until the water's surface temperature cools to the freezing point, and then below the freezing point—a process called supercooling. When this occurs, ice particles begin to form above the water's surface.

Though wind and currents typically interfere with the freezing process, ice can form on fast-flowing rivers. Ice crystals, called frazil, can appear as a thin layer of ice in a slow-moving river or as slush in faster currents. As ice forms, it slows down the velocity of the water. As long as air temperatures remain below freezing 32° F (o°C) near the surfaces of lakes and rivers, ice will thicken from the bottom layers up. In other words, the water is at its coldest when the freezing of its surface begins. As temperatures increase, however, ice on rivers and lakes decays due to the penetration of solar radiation.

The salinity of the world's oceans keeps them from freezing at the same temperature as bodies of freshwater. Salt ions disrupt the formation of ice crystals at the normal freezing temperature. However, in extreme polar areas, sea ice can form. There are three kinds of sea ice: landfast ice, which is attached to another surface or caught between icebergs; marine ice, which forms on the bottom of Antarctica's ice shelves; and pack ice, which drifts with currents and wind. As with ice on lakes and rivers, sea ice experiences changes throughout the seasons.

Sea ice serves several purposes, such as providing a habitat to ocean life. Arctic and Antarctic sea ice supports bacteria, algae, and fungi that live off the nutritive brine and, in turn, become a food source for organisms such as krill. Sea ice also provides a measure of climate control. Snow that comes to rest on the ice insulates the ocean water from colder air above, retaining heat that influences weather and helps to sustain ocean life.

Ice also forms on land as well as within bodies of water. Permafrost, which is moisture-laden ground that is frozen for two years or longer, covers about 20 percent of Earth's surface. It exists in climates where the average annual air temperature is 32° F (o°C) or colder. For permafrost to form, consistently low air temperatures draw the thermal energy from deep inside the earth, freezing water within the soil. Each winter, the layers of permafrost grow deeper until halted by the heat of Earth's core. It takes thousands of years for permafrost to embed itself hundreds of feet below the surface.

In summer, warmer air temperatures heat the top "active" layer of the ground, a region that is just a few meters thick. As permafrost freezes and thaws year after year, it produces large-scale geomorphic polygon patterns on the surface. Thawing permafrost can create small mounds or depressions in the earth, as well as larger, more substantial tunnels and caverns.

There are five types of ground ice found in permafrost: pore, segregated (Taber), foliated (wedge), pingo, and buried ice. Each type is the result of the way in which water is distributed in the ground. For instance, pingos are hills of soil and rock with an ice core; some have been measured at almost 200 feet (30 metres). Pingos form as groundwater freezes, causing the water and sediment to "heave" upward.

Permafrost dictates how hundreds of thousands of people live. Parts of Canada, Alaska, China, and Russia lie on permafrost. Crop and plant growth are difficult in these areas, at best. Construction and transportation companies must factor in the changes in the active permafrost layer when planning infrastructure. Engineers have managed to work around some problems. For example, parts of the Trans-Alaska Pipeline System run above ground, raised above the surface on lines of individual supports. However, permafrost remains a daunting obstacle; some buildings, roads, railroads, and pipes need annual replacement.

Large bodies of ice that form on land aboveground, due to the recrystallization of solid water such as snow, are called glaciers. Glaciers start with ice crystals. As snow falls, the individual flakes break down under the forces of wind, evaporation, and pressure from overlying layers of precipitation to forge small, hard grains of ice. As more snow falls, the layers become more densely packed, and ice grains become larger, rounder, and less likely to allow gas and liquid to penetrate the mass. For the glacier to grow, it needs to accumulate more mass through precipitation, condensation, and other processes than will be lost through erosion, via melting or runoff. The process of erosion is known as ablation.

Glaciers are among the largest moving objects on Earth, though typically their motion is so slow as to be imperceptible. They have the properties of ice, and therefore can suffer the same consequences caused by stress. Top layers can become brittle and crack, creating deep crevasses. Lower layers are more plastic, meaning they spread out to relieve the pressure of weight. Stress from the top layers may cause the glacier's base to slip over the bedrock beneath. Water at the base—sometimes supplied by leaking crevasses or friction at the base—facilitates glacial movement.

While a majority of the world's ice can be found in regions near the North and South poles, glaciers can grow elsewhere--including in the cold, mountainous terrain near the Equator. Mountain glaciers are especially susceptible to bed slips as gravity acts on their mass, pulling them down mountain slopes. Types of mountain glaciers include valley glaciers and ice fields. Piedmont glaciers are found in flat areas, fed by valley glaciers. Hanging glaciers cling precariously to mountains, sometimes resulting in avalanches.

When mountain glaciers advance over bedrock, they greatly alter the landscape. Their enormous mass can "absorb" terrain as the melting base refreezes around rocks, prying them out of the surface. As a glacier retreats, or melts, it may leave behind what is known as till, comprised of huge boulders, hills of smaller rocks, or tiny bits of "rock flour." Moraines, fjords, and arêtes are land formations molded by glacial movement.

The largest glaciers, called ice sheets, extend over a large area and flow in many directions. The Antarctic and Greenland ice sheets hold 99 percent of the world's glacier ice. The Antarctic ice sheet covers an area of roughly 5.3 million square miles (13.8 million square kilometres). Though much smaller than the Antarctic sheet, the Greenland ice sheet is still much larger than any mountain glacier.

Glacial flow on ice sheets generally drains from high elevations to basins that feed ice streams. Flow is very slow in the interior of ice sheets, perhaps inches each year, speeding up towards the outer edges, hundreds of feet per year. The flow may create ice shelves, which are thick slabs of an ice sheet that extend into the ocean. Ice shelves may be hammered by waves, tides, ocean currents, sea ice, and other icebergs. These stresses cause crevasses that eventually sever and calve, or break apart from the ice shelf, unleashing icebergs.

Antarctica produces the largest icebergs. Called tabular icebergs, they may have a freeboard (height above water) of over 150 feet (45 metres) and be up to 1,300 feet (395 metres) thick. Occasionally, some of these bergs measure over 150 miles (240 km) long. In contrast, most Arctic icebergs are calved from fast-flowing glaciers; the mountainous terrain stresses the glacial flow, creating more crevasses. Arctic icebergs generally are smaller than their Antarctic counterparts. Even ice islands, the largest of the Arctic icebergs, are only about 200 feet (60 metres) thick. All icebergs, regardless of their region of origin, are much larger than they appear above water.

All icebergs warm as soon as they break away from their parent glacier. Meltwater eats away at the glacier from the top, although it may refreeze as it seeps into the berg's cold core. Ocean waves also cause erosion. Icebergs may travel great distances, losing mass as they enter warmer waters. Some icebergs carry sediment and plant life, which help scientists identify the iceberg's origin.

As evidenced by the HMS *Titanic* tragedy of 1912, icebergs have historically been a danger to ships. Currently, ice patrols in the North Atlantic track icebergs and transmit reports to ships in the area. No such system is in place in the Southern Ocean. Ships, aircraft, and, most recently, satellites are used to keep track of icebergs that may invade shipping lanes. If icebergs need to be removed, they are usually towed. Special explosives that could fragment icebergs have not yet been perfected. Difficulties are compounded by the danger of affixing devices to capsize-prone icebergs.

Glaciologists believe that massive ice formations such as the Antarctic and Greenland ice sheets, offer a glimpse of Earth's past—and possibly its future. Snow that has fallen for millions of years has accumulated in layers of ice that, when extracted as cores and tested, reveal climatic data from prehistoric times. Air bubbles in glacial ice have confirmed growing amounts of carbon dioxide in the air. The increases in atmospheric carbon dioxide concentration is a contributing factor to global warming, which is an increase in Earth's near-surface air temperatures believed to be caused by human activities. Global warming has been cited as a reason for the decrease in Arctic sea ice; the near disappearance of the Ward Hunt Ice Shelf in Canada's Ellesmere Island, along with parts Antarctica's Larsen and Wilkins ice shelves; and the continued retreat of the world's mountain glaciers. Global warming also has been named a culprit in rising sea levels, although the amount of that increase due to glacial ablation is uncertain. Recent core samplings offer evidence that climate change can occur quite rapidly, perhaps as much as 9° to 13°F (5° to 7°C) within a single decade. In addition, it has been estimated that if all Earth's glaciers melted, sea levels would likely rise about 300 feet (90 metres) worldwide, and humans would lose three-fourths of Earth's freshwater to the ocean. Facts and statistics such as these only serve to underscore the tremendous power of ice.

CHAPTER 1
ICE ON
PLANET EARTH

Put simply, ice is a solid substance produced by the freezing of water vapour or liquid water. At temperatures below 0°C (32°F), water vapour develops into frost at ground level and snowflakes (each of which consists of a single ice crystal) in clouds. Below the same temperature, liquid water forms a solid as, for example, river ice, sea ice, hail, and ice produced commercially or in household refrigerators.

Ice occurs on the Earth's continents and surface waters in a variety of forms. Most notable are the continental glaciers (ice sheets) that cover much of Antarctica and Greenland. Smaller masses of perennial ice called ice caps occupy parts of Arctic Canada and other high-latitude regions, and mountain glaciers occur in more restricted areas, such as mountain valleys and the flatlands below. Other occurrences of ice on land include the different types of ground ice associated with permafrost—that is, permanently frozen soil common to very cold regions. In the oceanic waters of the polar regions, icebergs occur when large masses of ice break off from glaciers or ice shelves and drift away. The freezing of seawater in these regions results in the formation of sheets of sea ice known as pack ice. During the winter months, similar ice bodies form on lakes and rivers in many parts of the world.

THE WATER MOLECULE

Water is an extraordinary substance, anomalous in nearly all its physical and chemical properties and easily the most complex of all the familiar substances that are single-chemical compounds. Consisting of two atoms of hydrogen (H) and one atom of oxygen (O), the water molecule has the chemical formula H_2O. These three atoms are covalently bonded (i.e., their nuclei are linked by attraction to shared electrons) and form a specific structure, with the oxygen atom located between the two hydrogen atoms. The three atoms do not lie in a straight line, however. Instead, the hydrogen atoms are bent toward each other, forming an angle of about $105°$.

The three-dimensional structure of the water molecule can be pictured as a tetrahedron with an oxygen nucleus centre and four legs of high electron probability. The two legs in which the hydrogen nuclei are present are called bonding orbitals. Opposite the bonding orbitals and directed to the opposite corners of the tetrahedron are two legs of negative electrical charge. Known as the lone-pair orbitals, these are the keys to water's peculiar behaviour, in that they attract the hydrogen nuclei of adjacent water molecules to form what are called hydrogen bonds. These bonds are not especially strong, but because they orient the water molecules into a specific configuration, they significantly affect the properties of water in its solid, liquid, and gaseous states.

In the liquid state, most water molecules are associated in a polymeric structure — that is, chains of molecules connected by weak hydrogen bonds. Under the influence of thermal agitation, there is a constant breaking and reforming of these bonds. In the gaseous state, whether steam or water vapour, water molecules are largely independent of one another, and, apart from collisions, interactions between them are slight. Gaseous water, then, is largely monomeric — that is, consisting of single molecules—although there occasionally occur dimers (a union of two molecules) and even some trimers (a combination of three molecules).

In the solid state, at the other extreme, water molecules interact with one another strongly enough to form an ordered crystalline structure, with each oxygen atom collecting the four nearest of its neighbours and arranging them about itself in a rigid lattice. This structure results in a more open assembly, and hence a lower density, than the closely packed assembly of molecules in the liquid phase. For this reason, water is one of the few substances that is actually less dense in solid form than in the liquid state, dropping from 1,000 to 917 kg per cubic metre (62 to 57 pounds per cubic foot). It is the reason why ice floats rather than sinking, so that, during the winter, it develops as a sheet on the surface of lakes and rivers rather than sinking below the surface and accumulating from the bottom.

As water is warmed from the freezing point of 0 to 4°C (from 32 to 39°F), it contracts and becomes denser. This initial increase in density takes place because at 0°C a portion of the water consists of open-structured molecular arrangements similar to those of ice crystals. As the temperature increases, these structures break down and reduce their volume to that of the more closely packed polymeric structures of the liquid state. With further warming beyond 4°C, the water begins to expand in volume, along with the usual increase in intermolecular vibrations caused by thermal energy.

THE ICE CRYSTAL

At standard atmospheric pressure and at temperatures near 0°C (32°F), the ice crystal commonly takes the form of sheets or planes of oxygen atoms joined in a series of open hexagonal rings. The axis parallel to the hexagonal rings is termed the c-axis and coincides with the optical axis of the crystal structure.

When viewed perpendicular to the c-axis, the planes appear slightly dimpled. The planes are stacked in a laminar structure that occasionally deforms by gliding, like a deck of cards. When this gliding deformation occurs, the bonds between the layers break, and the hydrogen atoms involved in those bonds must become attached to different oxygen atoms. In doing so, they migrate within the lattice, more rapidly at higher temperatures. Sometimes they do not reach the usual arrangement of two hydrogen atoms connected by covalent bonds to each oxygen atom, so that some oxygen atoms have only one or as many as three hydrogen bonds. Such oxygen atoms become the sites of electrical charge. The speed of crystal deformation depends on these readjustments, which in turn are sensitive to temperature. Thus the mechanical, thermal, and electrical properties of ice are interrelated.

Hoarfrost and Rime

Hoarfrost is a deposit of ice crystals on objects exposed to the free air, such as grass blades, tree branches, or leaves. It is formed by direct condensation of water vapour to ice at temperatures below freezing and occurs when air is brought to its frost point by cooling. Hoarfrost is formed by a process analogous to that by which dew is formed on similar objects, except that, in the case of dew, the saturation point of the air mass is above freezing. The occurrence of temperatures below 0°C

(32°F) is not enough to guarantee the formation of hoarfrost. Additionally, the air must be initially damp enough so that when cooled it reaches saturation, and any additional cooling will cause condensation to occur. In the absence of sufficient moisture, hoarfrost does not form, but the water in the tissues of plants may freeze, producing the condition known as black frost.

Rime is described as a white, opaque, granular deposit of ice crystals formed on objects that are at a temperature below the freezing point. Rime occurs when supercooled water droplets (at a temperature lower than 0°C [32°F]) in fog come in contact with a surface that is also at a temperature below freezing; the droplets are so small that they freeze almost immediately upon contact with the object. Rime is common on windward upper slopes of mountains that are enveloped by supercooled clouds. These rime deposits take the form of long plumes of ice oriented into the direction of the wind and are called "frozen fog deposits," or "frost feathers." Rime is composed of small ice particles with air pockets between them; this structure causes its typical white appearance and granular structure. Because of the rapid freezing of each individual supercooled droplet, there is relatively poor cohesion between the neighbouring ice particles, and the deposits may easily be shattered or removed from objects they form on. Thus, rime is not normally a serious problem when it forms on the wings or other surfaces of aircraft.

MECHANICAL PROPERTIES

Like any other crystalline solid, ice subject to stress undergoes elastic deformation, returning to its original shape when the stress ceases. However, if a shear stress or force is applied to a sample of ice for a long time, the sample will first deform elastically and will then continue to deform plastically, with a permanent alteration of shape. This plastic deformation, or creep, is of great importance to the study of glacier flow. It involves two processes: intracrystalline gliding, in which the layers within an ice crystal shear parallel to each other without destroying the continuity of the crystal lattice, and recrystallization, in which crystal boundaries change in size or shape depending on the orientation of the adjacent crystals and the stresses exerted on them. The motion of dislocations—that is, of defects or disorders in the crystal lattice—controls the speed of plastic deformation. Dislocations do not move under elastic deformation.

The strength of ice, which depends on many factors, is difficult to measure. If ice is stressed for a long time, it deforms by plastic flow and has no yield point (at which permanent deformation begins) or ultimate strength. For short-term experiments with conventional testing machines, typical strength values in bars are 38 for crushing, 14 for bending, 9 for tensile, and 7 for shear.

THERMAL PROPERTIES

The heat of fusion (heat absorbed on melting of a solid) of water is 334 kilojoules per kilogram. The specific heat of ice at the freezing point is 2.04 kilojoules per kilogram per degree Celsius. The thermal conductivity at this temperature is 2.24 watts per metre kelvin.

Another property of importance to the study of glaciers is the lowering of the melting point due to hydrostatic pressure: 0.0074°C (0.013°F) per bar. Thus, for a glacier 300 metres (984 feet) thick, everywhere at the melting temperature, the ice at the base is 0.25°C (0.45°F) colder than at the surface.

OPTICAL PROPERTIES

Pure ice is transparent, but air bubbles render it some-what opaque. The absorption coefficient, or rate at which incident radiation decreases with depth, is about 0.1 cm^{-1} for snow and only 0.001 cm^{-1} or less for clear ice. Ice is weakly birefringent, or doubly refracting, which means that light is transmitted at different speeds in different crystallographic directions. Thin sections of snow or ice therefore can be conveniently studied under polarized light in much the same way that rocks are studied. The ice crystal strongly absorbs light in the red wavelengths, and thus the scattered light seen emerging from glacier crevasses and unweathered ice faces appears as blue or green.

ELECTROMAGNETIC PROPERTIES

The albedo, or reflectivity (an albedo of 0 means that there is no reflectivity), to solar radiation ranges from 0.5 to 0.9 for snow, 0.3 to 0.65 for firn, and 0.15 to 0.35 for glacier ice. At the thermal infrared wavelengths, snow and ice are almost perfectly "black" (absorbent), and the albedo is less than 0.01. This means that snow and ice can either absorb or radiate long-wavelength radiation with high efficiency. At longer electromagnetic wavelengths (microwave and radio frequencies), dry snow and ice are relatively transparent, although the presence of even small amounts of liquid water greatly modifies this property. Radio echo sounding (radar) techniques are now used routinely to measure the thickness of dry polar glaciers, even where they are kilometres in thickness, but the slightest amount of liquid water distributed through the mass creates great difficulties with the technique.

Ice Ages

Ice ages, which are also known as glacial ages, are any geologic period during which thick ice sheets cover vast areas of land. Such periods of large-scale glaciation may last several million years and drastically reshape surface features of entire continents. A number of major ice ages have occurred throughout Earth history. The earliest known took place during Precambrian time dating back more than 570 million years. The most recent periods of widespread glaciation occurred during the Pleistocene Epoch (about 2.6 million to 11,700 years ago).

A lesser, recent glacial stage called the Little Ice Age began in the 16th century and advanced and receded intermittently over three centuries in Europe and many other regions. Its maximum development was reached about 1750, at which time glaciers were more widespread on Earth than at any time since the last major ice age ended about 11,700 years ago.

CHAPTER 2

PERMAFROST

Permafrost is perennially frozen ground, a naturally occurring material with a temperature colder than 0°C (32°F) continuously for two or more years. Such a layer of frozen ground is designated exclusively on the basis of temperature. Part or all of its moisture may be unfrozen, depending on the chemical composition of the water or the depression of the freezing point by capillary forces. Permafrost with saline soil moisture, for example, may be colder than 0 °C for several years but contain no ice and thus not be firmly cemented. Most permafrost, however, is consolidated by ice.

Permafrost with no water, and thus no ice, is termed dry permafrost. The upper surface of permafrost is called the permafrost table. In permafrost areas, the surface layer of ground that freezes in the winter (seasonally frozen ground) and thaws in summer is called the active layer. The thickness of the active layer depends mainly on the moisture content, varying from less than a foot in thickness in wet, organic sediments to several feet in well-drained gravels.

Permafrost forms and exists in a climate where the mean annual air temperature is 0°C (32°F) or colder. Such a climate is generally characterized by long, cold winters with little snow and short, relatively dry, cool summers. Permafrost, therefore, is widespread in the Arctic, sub-arctic, and Antarctica. It is estimated to underlie 20 percent of the world's land surface.

Thawed surface of the permafrost on the tundra in summer, Taymyr Peninsula, Siberia. © John Hartley/NHPA

THE ORIGIN AND STABILITY OF PERMAFROST

The strength and thickness of permafrost in a given region depends on variations in air and ground temperature, as well as geothermal heating. The depth of the frozen ground waxes and wanes according to seasonal cycles. As the process of climate change brings warmer conditions to many cold regions, some areas of frozen ground are melting in order to reach a new equilibrium with the present climate.

AIR TEMPERATURE AND GROUND TEMPERATURE

In areas where the mean annual air temperature becomes colder than 0°C (32°F), some of the ground frozen in the winter will not be completely thawed in the summer. Therefore, a layer of permafrost will form and continue to grow downward gradually each year from the seasonally frozen ground. The permafrost layer will become thicker each winter, its thickness controlled by the thermal balance between the heat flow from the Earth's interior and that flowing outward into the atmosphere. This balance depends on the mean annual air temperature and the geothermal gradient. The average geothermal gradient is an increase of 1°C (1.8°F) for every 30 to 60 metres (about 100 to 200 feet) of depth. Eventually the thickening permafrost layer reaches an equilibrium depth at which the amount of geothermal heat reaching the permafrost is on the average equal to that lost to the atmosphere. Thousands of years are required to attain a state of equilibrium where permafrost is hundreds of feet thick.

The annual fluctuation of air temperature from winter to summer is reflected in a subdued manner in the upper few metres of the ground. This fluctuation diminishes rapidly with depth, being only a few degrees at 7.5 metres (about 25 feet), and is barely detectable at 15 metres (about 50 feet). The level of zero amplitude, at which fluctuations are hardly detectable, is 9 to 15 metres (30 to 50 feet). If the permafrost is in thermal equilibrium, the temperature at the level of zero amplitude is generally regarded as the minimum temperature of the permafrost. Below this depth the temperature increases steadily under the influence of heat from the Earth's interior. The temperature of permafrost at the depth of minimum annual seasonal change varies from near 0°C at the southern limit of permafrost to -10°C (14°F) in northern Alaska and -13°C (9°F) in northeastern Siberia.

As the climate becomes colder or warmer, but maintaining a mean annual temperature colder than 0°C, the temperature of the permafrost correspondingly rises or declines, resulting in changes in the position of the base of permafrost. The position of the top of permafrost will be lowered by thawing when the climate warms to a mean annual air temperature warmer than 0°C. The rate at which the base or top of permafrost is changed depends not only on the amount of climatic fluctuation but also on the amount of ice in the ground and the composition of the ground, conditions that in part control the geothermal gradient. If the geothermal gradient is known and if the surface temperature remains stable for a long period of time, it is, therefore, possible to predict from a knowledge of the mean annual air temperature the thickness of permafrost in a particular area that is remote from bodies of water.

CLIMATIC CHANGE

Permafrost is the result of present climate. Many temperature profiles show, however, that permafrost is not in equilibrium with present climate at the sites of measurement. Some areas show, for example, that climatic warming since the last third of the 19th century has caused a warming of the permafrost to a depth of more than 100 metres (about 330 feet). In such areas much of the permafrost is a product of a colder, former climate.

The distribution and characteristics of subsea permafrost point to a similar origin. At the height of the glacial epoch, especially about 20,000 years ago, most of the continental shelf in the Arctic Ocean was exposed to polar climates for thousands of years. These climates caused cold permafrost to form to depths of more than

700 metres (about 2,300 feet). Subsequently, within the past 10,000 years, the Arctic Ocean rose and advanced over a frozen landscape to produce a degrading relict subsea permafrost. The perennially frozen ground is no longer exposed to a cold atmosphere, and the salt water has caused a reduction in strength and consequent melting of the ice-rich permafrost (which is bonded by freshwater ice). The temperature of subsea permafrost, near -1°C (30°F), is no longer as low as it was in glacial times and is therefore sensitive to warming from geothermal heat and to the encroaching activities of humans.

It is thought that permafrost first occurred in conjunction with the onset of glacial conditions about three million years ago, during the late Pliocene Epoch. In the subarctic at least, most permafrost probably disappeared during interglacial times and reappeared in glacial times. Most existing permafrost in the subarctic probably formed in the cold (glacial) period of the past 100,000 years.

PERMAFROST DISTRIBUTION IN THE NORTHERN HEMISPHERE

Permafrost is widespread in the northern part of the Northern Hemisphere, where it occurs in 85 percent of Alaska and 55 percent of Russia and Canada. In contrast, in the Southern Hemisphere, permafrost is probably found across all of Antarctica. Permafrost is more widespread and extends to greater depths in the north than in the south. It is 1,500 metres (5,000 feet) thick in northern Siberia, 740 metres (about 2,400 feet) thick in northern Alaska, and thins progressively toward the south.

Most permafrost can be differentiated into two broad zones, the continuous and the discontinuous, referring to the lateral continuity of permafrost. In the continuous zone of the far north, permafrost is nearly everywhere present except under the lakes and rivers that do not freeze to the bottom. The discontinuous zone includes numerous permafrost-free areas that increase progressively in size and number from north to south. Near the southern boundary, only rare patches of permafrost have been found to exist.

In addition to its widespread occurrence in the Arctic and subarctic areas of the Earth, permafrost also exists at lower latitudes in areas of high elevation. This type of perennially frozen ground is called Alpine permafrost. Although data from high plateaus and mountains are scarce, measurements taken below the active surface layer indicate zones where temperatures of 0°C or colder persist for two or more years. The largest area of Alpine permafrost is in western China, where 1.5 million square km (580,000 square miles) of permafrost are known to exist. In the contiguous United States, Alpine permafrost is limited to about 100,000 square km (38,600 square miles) in the high mountains of the west. Permafrost occurs at elevations as low as 2,500 metres (8,200 feet) in the northern states and at about 3,500 metres (about 11,500 feet) in Arizona.

A unique occurrence of permafrost—one that has no analogue on land—lies under the Arctic Ocean, on the northern continental shelves of North America and Eurasia. This is known as subsea or offshore permafrost.

THE LOCAL THICKNESS OF PERMAFROST

The thickness and areal distribution of permafrost are directly affected by snow and vegetation cover, topography, bodies of water, the interior heat of the Earth, and the temperature of the atmosphere.

THE EFFECTS OF CLIMATE

The most conspicuous change in thickness of permafrost is related to climate. At Barrow, Alaska, the mean annual air temperature is -12°C (10°F), and the thickness is 400 metres (about 1,300 feet). At Fairbanks, Alaska, in the discontinuous zone of permafrost in central Alaska, the mean annual air temperature is -3°C (27°F), and the thickness is about 90 metres (about 300 feet). Near the southern border of permafrost, the mean annual air temperature is about 0 or -1°C, and the perennially frozen ground is only a few feet thick.

If the mean annual air temperature is the same in two areas, the permafrost will be thicker where the conductivity of the ground is higher and the geothermal gradient is less. A. H. Lachenbruch of the U.S. Geological Survey reported an interesting example from northern Alaska. The mean annual air temperatures at Cape Simpson and Prudhoe Bay are similar, but permafrost thickness is 275 metres (900 feet) at Cape Simpson and about 650 metres (about 2,100 feet) at Prudhoe Bay because rocks at Prudhoe Bay are more siliceous and have a higher conductivity and a lower geothermal gradient than rocks at Cape Simpson.

THE EFFECTS OF WATER BODIES

Bodies of water, lakes, rivers, and the sea have a profound effect on the distribution of permafrost. A deep lake that does not freeze to the bottom during the winter will be underlain by a zone of thawed material. If the minimum horizontal dimension of the deep lake is about twice as much as the thickness of permafrost nearby, there probably exists an unfrozen vertical zone extending all the way to the bottom of permafrost. Such thawed areas extending all the way through permafrost are widespread under rivers and sites of recent rivers in the discontinuous zone of permafrost and under major, deep rivers in the far north. Under the wide floodplains of rivers in the subarctic, the permafrost is sporadically distributed both laterally and vertically. Small, shallow lakes that freeze to the bottom each winter are underlain by a zone of thawed material, but the thawed zone does not completely penetrate permafrost except near the southern border of permafrost.

THE EFFECTS OF SOLAR RADIATION, VEGETATION, AND SNOW COVER

In as much as south-facing hillslopes receive more incoming solar energy per unit area than other slopes, they are warmer. Permafrost is generally absent on these in the discontinuous zone and is thinner in the continuous zone. The main role of vegetation in permafrost areas is to shield perennially frozen ground from solar energy. Vegetation is an excellent insulating medium and removal or disturbance of it, either by natural processes or by humans, causes thawing of the underlying permafrost. In the continuous zone the permafrost table may merely be lowered by the disturbance of vegetation, but in a discontinuous zone permafrost may be completely destroyed in certain areas.

Snow cover also influences heat flow between the ground and the atmosphere and therefore affects the distribution of permafrost. If the net effect of timely snowfalls is to prevent heat from leaving the ground in the cold winter, permafrost becomes warmer. Actually, local differences in vegetation and snowfall in areas of

Greenery, such as the vegetation surrounding this permafrost lake in Canada, absorbs heat from the sun and insulates the frozen ground underneath. Shutterstock.com

thin and warm permafrost are critical for the formation and existence of the perennially frozen ground. Permafrost is not present in areas of the world where great snow thicknesses persist throughout most of the winter.

TYPES OF GROUND ICE

The ice content of permafrost is probably the most important feature of permafrost affecting human life in the north. Ice in the perennially frozen ground exists in various sizes and shapes and has definite distribution characteristics. The forms of ground ice can be grouped into five main types: pore ice, segregated (or Taber) ice, foliated (or wedge) ice, pingo ice, and buried ice.

Pore ice, which fills or partially fills pore spaces in the ground, is formed by pore water freezing in situ with no addition of water. The ground contains no more water in the solid state than it could hold in the liquid state.

Segregated, or Taber, ice includes ice films, seams, lenses, pods, or layers generally 0.15 to 13 cm (0.06 to 5 inches) thick that grow in the ground by drawing in water as the ground freezes. Small ice segregations are the least spectacular but one of the most extensive types of ground ice. Engineers and geologists interested in ice growth and its effect on engineering structures have studied them considerably. Such observers generally accept the principle of bringing water to a growing ice crystal, but they do not completely agree as to the mechanics of the processes. Pore ice and Taber ice occur both in seasonally frozen ground and in permafrost.

Foliated ground ice, or wedge ice, is the term for large masses of ice growing in thermal contraction cracks in permafrost, while pingo ice is clear, or relatively clear, and occurs in permafrost more or less horizontally or in lens-shaped masses. Such ice originates from groundwater under hydrostatic pressure.

Buried ice in permafrost includes buried sea, lake, and river ice and recrystallized snow, as well as buried blocks of glacier ice in permafrost climate.

World estimates of the amount of ice in permafrost vary from 200,000 to 500,000 cubic km (49,000 to 122,000 cubic miles), or less than 1 percent of the total volume of the Earth. It has been estimated that 10 percent by volume of the upper 3 metres (10 feet) of permafrost on the northern Coastal Plain of Alaska is composed of foliated ground ice (ice wedges). Taber ice is the most extensive type of ground ice, and in places it represents 75 percent of the ground by volume. It is calculated that the pore and Taber ice content in the depth between 0.5 and 3 metres (2 and 10 feet)—the surface to 0.5 metre is seasonally thawed—is 61 percent by volume, and between 3 and 9 metres (10 and 30 feet) it is 41 percent. The total amount of pingo ice is less than 0.1 percent of the permafrost. The total ice content in the permafrost of the Arctic Coastal Plain of Alaska is estimated to be 1,500 cubic km (360 cubic miles), and below 9 metres (30 feet) most of that is present as pore ice.

ICE WEDGES

The most conspicuous and controversial type of ground ice in permafrost is that formed in large ice wedges or masses with parallel or subparallel foliation structures. Most foliated ice masses occur as wedge-shaped, vertical, or inclined sheets or dikes 0.025 to 3 metres (0.1 to 10 feet) wide and 0.3 to 9 metres (1 to 30 feet) high when viewed in transverse cross section. Some masses seen on the face of frozen cliffs may appear as horizontal bodies a few centimetres to 3 metres (10 feet) in thickness and 0.3 to 14 metres (1 to 46 feet) long, but the true shape of these ice wedges can be seen only in three dimensions. Ice wedges are parts of polygonal networks of ice enclosing cells of frozen ground 3 to 30 metres (10 to 100 feet) or more in diameter.

The origin of ground ice was first studied in Siberia, and discussions in print of the origin of large ground-ice masses in perennially frozen ground of North America have gone on since Otto von Kotzebue recorded ground ice in 1816 at a spot now called Elephant's Point in Eschscholtz Bay of Seward Peninsula. The theory for the origin of ice wedges now generally accepted is the thermal contraction theory that, during the cold winter, polygonal thermal contraction cracks, a centimetre or two wide and a few metres deep, form in the frozen ground. Then, when water from the melting snow runs down these tension cracks and freezes in early spring, a vertical vein of ice is produced that penetrates into permafrost. When the permafrost warms and re-expands during the following summer, horizontal compression produces upturning of the frozen sediment by plastic deformation. During the next winter, renewed thermal tension reopens the vertical ice-cemented crack, which may be a zone of weakness. Another increment of ice is added in the spring when meltwater again enters and freezes. Over the years, the vertical wedge-shaped mass of ice is produced.

ACTIVE WEDGES, INACTIVE WEDGES, AND ICE-WEDGE CASTS

Ice wedges may be classified as active, inactive, and ice-wedge casts. Active ice wedges are those that are actively growing. The wedge may not crack every year, but during many or most years cracking does occur, and an increment of ice is added. Ice wedges require a much more rigorous climate to grow than does permafrost. The permafrost table must be chilled to -15 to -20°C (5 to -4°F) for contraction cracks to form. On the average, it is assumed that ice wedges generally grow in a climate where the mean annual air temperature is -6 or -8°C (21 or 18°F) or colder. In regions with a general mean annual temperature only slightly warmer than -6°C (21°F), ice wedges occasionally form in restricted cold microclimate areas or during cold periods of a few years' duration.

The area of active ice wedges appears to roughly coincide with the continuous permafrost zone. From north to south across the permafrost area in North America, a decreasing number of wedges crack frequently. The line dividing zones of active and inactive ice wedges is arbitrarily placed at the position where it is thought most wedges do not frequently crack.

Inactive ice wedges are those that are no longer growing. The wedge does not crack in winter and, therefore, no new ice is added. A gradation between active ice wedges and inactive ice wedges occurs in those wedges that crack rarely. Inactive ice wedges have no ice seam or crack extending from the wedge upward to the surface in the spring. The wedge top may be flat, especially if thawing has lowered the upper surface of the wedge at some time in the past.

Ice wedges in the world are of several ages, but none appear older than the onset of the last major cold period, about 100,000 years ago. Wedges dated by radiocarbon analyses range from 3,000 to 32,000 years in age.

In many places in the now temperate latitudes of the world, in areas of past permafrost, ice wedges have melted, and resulting voids have been filled with sediments collapsing from above and the sides. These ice-wedge casts are important as paleoclimatic indicators and indicate a climate of the past with at least a mean annual air temperature of -6 or -8°C (21 or 18°F) or colder.

SURFACE MANIFESTATIONS OF PERMAFROST AND SEASONALLY FROZEN GROUND

Many distinctive surface manifestations of permafrost exist in the Arctic and subarctic, including such geomorphic features as polygonal ground, thermokarst phenomena, and pingos. In addition to the above, there are many features caused in large part by frost action that are common in but not restricted to permafrost areas, such as solifluction (soil flowage) and frost-sorted patterned ground.

POLYGONAL GROUND

One of the most widespread geomorphic features associated with permafrost is the microrelief pattern on the surface of the ground generally called polygonal ground, or tundra polygons. This pattern, which covers thousands of square miles of the Arctic and less in the subarctic, is caused by an intersecting network of shallow troughs delineating polygons 3 to 30 metres (10 to 100 feet) in diameter. The troughs are underlain by more or less vertical ice wedges 0.6 to 3 metres (2 to 10 feet) across on the top that are joined together in a honeycomb network. These large-scale polygons should not be confused with the small-scale polygons or patterned ground produced by frost sorting.

The ice-wedge polygons may be low-centred or high-centred. Upturning of strata adjacent to the ice wedge may make a ridge of ground on the surface on each side of the wedge, thus enclosing the polygons. Such polygons are lower in the centre and are called low-centre polygons or raised-edge polygons and may contain a pond in the centre. Low-centre, or raised-edge, polygons indicate that ice wedges are actually growing and that the sediments are being actively upturned. If erosion, deposition, or thawing is more prevalent than the up-pushing of the sediments along the side of the wedge or if the material being pushed up cannot maintain itself in a low ridge, the low ridges will be absent, and there may be either no polygons at the surface or the polygons may be higher in the centre than the troughs over the ice wedges that enclose them. Both high-centre and low-centre tundra polygons are widespread in the polar areas and are good indicators of the presence of foliated ice masses; care must be taken, however, to demonstrate that the pattern is not a relic and an indication of ice-wedge casts.

In many parts of the temperate latitudes of Asia, Europe, and North America, incompletely developed or poorly developed polygonal ground occurs on the same scale as in the Arctic. These large-scale polygons in the nonpermafrost areas are excellent evidence of the former extent of permafrost and ice wedges in the past glacial period.

In many areas of the continuous permafrost zone surface, drainage follows the troughs of the polygons (tops of the ice wedges). At ice wedge junctions or elsewhere, melting may occur to form small pools. The joining of these small pools by a stream causes the pools to resemble beads on a string, a type of stream form called beaded drainage. Such drainage indicates the presence of perennially frozen, fine-grained sediments cut by ice wedges.

THERMOKARST FORMATIONS

The thawing of permafrost creates thermokarst topography, an uneven surface that contains mounds, sinkholes, tunnels, caverns, and steep-walled ravines caused by the melting of ground ice. The hummocky ground surface resembles karst topography in limestone areas. Thawing may result from artificial or natural removal of vegetation or from a warming climate.

Thawed depressions filled with water (thaw lakes, thermokarst lakes, cave-in lakes) are widespread in permafrost areas, especially in those underlain with perennially frozen silt. They may occur on hillsides or even on hilltops and are good indicators of ice-rich permafrost. Locally, deep thermokarst pits 6 metres (20 feet) deep and 9 metres (30 feet) across may form as ground ice melts. These openings may exist as undetected caverns for many years before the roof collapses. Such collapses in agricultural or construction areas are real dangers. Thermokarst mounds are polygonal or circular hummocks 3 to 15 metres (10 to 50 feet) in diameter and 0.3 to 2.5 metres (1 to 8 feet) high that are formed as a polygonal network of ice melts and leaves the inner-ice areas as mounds.

Pingos

The most spectacular landforms associated with permafrost are pingos, small ice-cored circular or elliptical hills of frozen sediments or even bedrock, 3 to more than 60

metres (10 to more than 200 feet) high and 15 to 450 metres (50 to 1,500 feet) in diameter. Pingos are widespread in the continuous permafrost zone and are quite conspicuous because they rise above the tundra, and are much less conspicuous in the forested area of the discontinuous permafrost zone. Pingos are generally cracked on top with summit craters formed by melting ice.

Pingos rising up from the Mackenzie River Delta in northwest Canada. The result of an upheaval of frozen sediment, pingos are common in permafrost areas such as the Canadian tundra. Steve McCutcheon/Visuals Unlimited, Inc.

There are two types of pingos, based on origin. The closed-system type forms in level areas when unfrozen groundwater in a thawed zone becomes confined on all sides by permafrost, freezes, and heaves the frozen overburden to form a mound. This type is larger and occurs mainly in tundra areas of continuous permafrost. The open-system type is generally smaller and forms on slopes when water beneath or within the permafrost penetrates the permafrost under hydrostatic pressure. A hydrolaccolith (water mound) forms and freezes, heaving the overlying frozen and unfrozen ground to produce a mound.

Present pingos are apparently the result of postglacial climate and are less than 4,000–7,000 years old. Pingos were present in now temperate latitudes during the latest glacial epoch and are now represented by low circular ridges enclosing bogs or lowlands.

Near the southern border of permafrost occur palsas, low hills and knobs of perennially frozen peat about 1.5 to 6 metres (5 to 20 feet) high, evidently forming with accumulation of peat and segregation of ice.

PATTERNED GROUND

Intense seasonal frost action, repeated freezing and thawing throughout the year, produces small-scale patterned ground. Repetitive freezing and thawing tends to stir and sort granular sediments, thus forming circles, stone nets, and polygons a few centimetres to 6 metres (20 feet) in diameter. The coarse cobbles and boulders form the outside of the ring and the finer sediments occur in the centre. The features require a rigorous climate with some fine-grained sediments and soil moisture, but they do not necessarily need underlying permafrost. Permafrost, however, forms an impermeable substratum that keeps the soil moisture available for frost action. On gentle slopes the stone nets may be distorted into garlands by downslope movement or, if the slope is steep, into stone stripes about half a metre (1.6 feet) wide and 30 metres (100 feet) long.

SOIL FLOW

In areas underlain by an impermeable layer (seasonally frozen ground or perennially frozen ground), the active layer is often saturated with moisture and is quite mobile. The progressive downslope movement of saturated detrital material under the action of gravity and working in conjunction with frost action is called solifluction. This material moves in a semifluid condition and is manifested by lobelike and sheetlike flows of soil on slopes. The lobes are up to 30 metres (100 feet) wide and have a steep front 0.3 to 1.5 metres (1 to 5 feet) high.

An outstanding feature of solifluction is the mass transport of material over low-angle slopes. Solifluction deposits are widespread in polar areas and consist of a blanket 0.3 to 1.8 metres (1 to 6 feet) thick of unstratified or poorly stratified, unsorted, heterogeneous, till-like detrital material of local origin. In many areas the terrain is characterized by relatively smooth, round hills and slopes with well-defined to poorly defined solifluction lobes or terraces. If the debris is blocky and angular and fine material is absent, the lobes are poorly developed or absent. Areas in which solifluction lobes are well formed lie almost entirely above or beyond the forest limit.

In many areas the frost-rived debris contains few fine materials and little water and consists of angular fragments of well-jointed, resistant rock. Under such circumstances, solifluction lobes do not often occur, but instead striking sheets or streams of angular rubble form. These are called rock streams or rubble sheets.

THE STUDY OF PERMAFROST

Although the existence of permafrost had been known to the inhabitants of Siberia for centuries, scientists of the Western world did not take seriously the isolated reports of a great thickness of frozen ground existing under northern forest and grasslands until 1836. Then the Russian naturalist Alexander Theodor von Middendorff measured temperatures to depths of approximately 100 metres (330 feet) of permafrost in the Shargin shaft, an unsuccessful well dug for the governor of the Russian-Alaskan Trading Company, at Yakutsk, and estimated that the permafrost was 215 metres (about 700 feet) thick. Since the late 19th century, Russian scientists and engineers have actively studied permafrost and applied the results of their learning to the development of Russia's north.

In a similar way, prospectors and explorers were aware of permafrost in the northern regions of North America for many years, but it was not until after World War II that systematic studies of perennially frozen ground were undertaken by scientists and engineers in the United States and Canada. Since exploitation of the great petroleum resources on the northern continental shelves began in earnest in the 1970s, investigations into subsea permafrost have progressed even more rapidly than have studies of permafrost on land.

Alpine permafrost studies had their beginning in the study of rock glaciers in the Alps of Switzerland. Although ice was known to exist in rock glaciers, it was not until after World War II that investigation by geophysical methods clearly demonstrated slow movement of perennial ice—that is, permafrost. In the 1970s and 1980s, detailed geophysical work and temperature and borehole examination of mountain permafrost began in Russia,

China, and Scandinavia, especially with regard to construction in high mountain and plateau areas.

PROBLEMS POSED BY PERMAFROST

Permafrost poses unique challenges to those who wish to develop Earth's polar regions. Thawing, subsidence, frost action, and freezing can wreak havoc on infrastructure and permanent constructions that disturb the frozen ground. To maintain the stability of the ground (thus reducing the maintenance costs of buildings and roads), construction activities must take these phenomena into account before buildings are erected and transportation lines are laid out. Often, buildings, pipes, and electrical lines are placed on piles above the ground, while roads remain unpaved.

PERMAFROST ENGINEERING

Development of the north demands an understanding of and the ability to cope with problems of the environment dictated by permafrost. Although the frozen ground hinders agricultural and mining activities, the most dramatic, widespread, and economically important examples of the influence of permafrost on life in the north involve construction and maintenance of roads, railroads, airfields, bridges, buildings, dams, sewers, and communication lines. Engineering problems are of four fundamental types: (1) those involving thawing of ice-rich permafrost and subsequent subsidence of the surface under unheated structures such as roads and airfields; (2) those involving subsidence under heated structures; (3) those resulting from frost action, generally intensified by poor drainage caused by permafrost; and (4) those involved only with the temperature of permafrost that causes buried sewer, water, and oil lines to freeze.

A thorough study of the frozen ground should be part of the planning of any engineering project in the north. It is generally best to attempt to disturb the permafrost as little as possible in order to maintain a stable foundation for engineering structures, unless the permafrost is thin; then, it may be possible to destroy the permafrost. The method of construction preserving the permafrost has been called the passive method; alternately, the destroying of permafrost is the active method.

Because thawing of permafrost and frost action are involved in almost all engineering problems in polar areas, it is advisable to consider these phenomena generally. The delicate thermal equilibrium of permafrost is disrupted when the vegetation, snow cover, or active layer is compacted. The permafrost table is lowered, the active layer is thickened, and considerable ice is melted. This process lowers the surface and provides (in summer) a wetter active layer with less bearing strength. Such disturbance permits a greater penetration of summer warming. It is common procedure to place a fill, or pad, of gravel under engineering works. Such a fill generally is a good conductor of heat and, if thin, may cause additional thawing of permafrost. The fill must be made thick enough to contain the entire amplitude of seasonal temperature variation—in other words, thick enough to restrict the annual seasonal freezing and thawing to the fill and the compacted active layer. Under these conditions no permafrost will thaw. Such a procedure is quite feasible in the Arctic, but in the warmer subarctic it is impractical because of the enormous amounts of fill needed. Under a heated building, profound thawing may occur more rapidly than under roads and airfields.

Frost action, the freezing and thawing of moisture in the ground, has long been known to seriously disrupt and destroy structures in both polar and temperate latitudes.

In the winter the freezing of ground moisture produces upward displacement of the ground (frost heaving), and in the summer excessive moisture in the ground brought in during the freezing operation causes loss of bearing strength. Frost action is best developed in silt-sized and silty clay-sized sediments in areas of rigorous climate and poor drainage. Polar latitudes are ideal for maximum frost action because most lowland areas are covered by fine-grained sediments, and the underlying permafrost causes poor drainage.

DEVELOPMENT IN PERMAFROST AREAS

Piles are used to support many, if not most, structures built on ice-rich permafrost. In regions of cold winters, many pile foundations in the ground are subject to seasonal freezing and, therefore, possibly subject to the damaging effect of frost heaving, which tends to displace the pile upward and thus to disturb the foundation of the structure. The displacement of piling is not limited to the far north, though maximum disturbance probably is encountered most widely in the subarctic. Expensive maintenance and sometimes complete destruction of bridges, school buildings, military installations, pipelines, and other structures have resulted from failure to understand the principles of frost heaving of piling.

A remarkable construction achievement in a permafrost environment is the Trans-Alaska Pipeline System. Completed in 1977, this 1,285-km-long (800-mile-long), 122-cm-diameter (48-inch-diameter) pipeline transports crude oil from Prudhoe Bay to an ice-free port at Valdez. The pipeline was originally designed for burial along most of the route. However, because the oil is transported at 70° to 80°C (158° to 176°F), such an installation would have thawed the adjacent permafrost, causing liquefaction, loss

A length of the Trans-Alaska Pipeline near Squirrel Creek, Alaska. The pipeline was designed and built with the preservation of large areas of permafrost in mind. George F. Herben/National Geographic/Getty Images

of bearing strength, and soil flow. To prevent destruction of the pipeline, about half of the line (615 km [380 miles]) is elevated onto beams held up by vertical support members. The pipeline safely discharges its heat into the air, while frost heaving of the 120,000 vertical support members is prevented by freezing them firmly into the permafrost through the use of special heat-radiating thermal devices.

Highways in polar areas are relatively few and mainly unpaved. They are subject to subsidence by thawing of permafrost in summer, frost heaving in winter, and loss of bearing strength on fine-grained sediments in summer. Constant grading of gravel roads permits maintenance of a relatively smooth highway. Where the road is paved over ice-rich permafrost, the roadway becomes rough and is much more costly to maintain than are unpaved roads. Many of the paved roads in polar areas have required resurfacing two or three times in a 10-year period.

Railroads particularly have serious construction problems and require costly upkeep in permafrost areas because of the necessity of maintaining a relatively low gradient and the subsequent location of the roadbed in ice-rich lowlands that are underlain with perennially frozen ground. The Trans-Siberian Railroad, the Alaska Railroad, and some Canadian railroads in the north are locally underlain by permafrost with considerable ground ice. As the large masses of ice melt each summer, constant maintenance is required to level these tracks. In winter, extensive maintenance is also required to combat frost heaving when local displacements of 2.5 to 35 cm (1 to 14 inches) occur in roadbeds and bridges.

Permafrost affects agricultural developments in many parts of the discontinuous permafrost zone. Its destructive effect on cultivated fields in both Russia and North America results from the thawing of large masses of ice in the permafrost. If care is not exercised in selecting areas to be cleared for cultivation, thawing of the permafrost may necessitate abandonment of fields or their reduction to pasturage.

One of the most active and exciting areas of permafrost engineering is in subsea permafrost. Knowledge of the distribution, type, and water or ice content of subsea permafrost is critical for planning petroleum exploration, locating production structures, burying pipelines, and driving tunnels beneath the seabed. Furthermore, the temperature of the seabed must be known in order to predict potential sites of accumulation of gas hydrates or areas in which groundwater or artesian pressures are likely. In addition, knowledge of the distribution of subsea permafrost permits a thorough interpretation of regional geologic history.

CHAPTER 3
ICE IN LAKES AND RIVERS

S heets or stretches of ice also form on the surface of lakes and rivers when the temperature drops below freezing (0°C [32°F]). The nature of the ice formations may be as simple as a floating layer that gradually thickens, or it may be extremely complex, particularly when the water is fast-flowing.

Much of the world experiences weather well below the freezing point, and in these regions ice forms annually in lakes and rivers. About half the surface waters of the Northern Hemisphere freeze annually. In warmer climates, waters may freeze only occasionally during periods of unusual cold, and in extremely cold areas of the world, such as Antarctica, lakes may have a permanent ice cover.

THE SEASONAL CYCLE

In most regions where ice occurs, the formation is seasonal in nature. An initial ice cover forms some time after the average daily air temperature falls below the freezing point. The ice cover thickens through the winter period and melts and decays as temperatures warm in the spring. During the formation and thickening periods, energy flows out of the ice cover, and, during the decay period, energy flows into the ice cover. This flow of energy consists of two basic modes of energy exchange: the radiation of long-wavelength and short-wavelength electromagnetic energy (i.e., infrared and ultraviolet light) and the transfer of heat energy associated with evaporation and condensation, with convection between the air and the surface, and (to a lesser extent) with precipitation falling on the surface.

While radiation transfers are important, the dominant energy exchange in ice formation and decay is the heat transfer associated with evaporation and condensation and with turbulent convection—the latter being termed the sensible transfer. Since these transfers of heat are driven by the difference between air temperature and surface temperature, the extent and duration of ice covers more or less coincide with the extent and duration of average air temperatures below the freezing point (with a lag in the autumn due to the cooling of the water from its summer heating and a lag in the spring due to the melting of ice formed over the winter).

As a general rule, small lakes freeze over earlier than rivers, and ice persists longer on lakes in the spring. Where there are sources of warm water—for example, in underground springs or in the thermal discharges of industrial power plants—this pattern may be disrupted, and water may be free of ice throughout the winter. In addition, in very deep lakes the thermal reserve built up during summer heating may be too large to allow cooling to the freezing point, or the action of wind over large fetches may prevent a stable ice cover from forming.

ICE FORMATION IN LAKES

The setting for the development of ice cover in lakes is the annual evolution of the temperature structure of lake water. In most lakes during the summer, a layer of warm water of lower density lies above colder water below. In late summer, as air temperatures fall, this top layer begins to cool. After it has cooled and has reached the same density as the water below, the water column becomes isothermal (i.e., there is a uniform temperature at all depths). With further cooling, the top water becomes even denser and plunges, mixing with the water below, so that the lake continues to be isothermal but at ever colder temperatures. This process continues until the temperature drops to that of the maximum density of water (about 4°C, or 39°F). Further cooling then results in expansion of the space between water molecules, so that the water becomes less dense.

This change in density tends to create a new stratified thermal structure, this time with colder, lighter water on top of the warmer, denser water. If there is no mixing of the water by wind or currents, this top layer will cool to the freezing point (0°C, or 32°F). Once it is at the freezing point, further cooling will result in ice formation at the surface. This layer of ice will effectively block the exchange of energy between the cold air above and the warm water below; therefore, cooling will continue at the surface, but, instead of dropping the temperature of the water below, the heat losses will be manifested in the production of ice.

The simple logic outlined above suggests that water at some depth in lakes during the winter will always be at 4°C (39°F), the temperature of maximum density, and indeed this is often the case in smaller lakes that are protected from the wind. The more usual scenario, however, is that wind mixing continues as the water column cools below 4°C, thereby overcoming the tendency toward density stratification. Between 4 and 0°C (39 and 32°F), for example, the density difference might be only 0.13 kilogram per cubic metre (3.5 ounces per cubic yard). Eventually some particular combination of cold air temperature, radiation loss, and low wind allows a first ice cover to form and thicken sufficiently to withstand wind forces that may break it up. As a result, even in fairly deep lakes the water temperature beneath the ice is usually somewhere below

As temperatures cool, lakes begin to freeze over from the top down. Water at greater depths tends to retain its liquid form, protected from freezing by the layer of ice at the surface. Shutterstock.com

4°C and quite often closer to 0°C. The temperature at initial ice formation may vary from year to year depending on how much cooling has occurred before conditions are right for the first initial cover to form and stabilize. In some large lakes, such as Lake Erie in North America, wind effects are so great that a stable ice cover rarely forms over the entire lake, and the water is very near 0°C throughout the winter.

NUCLEATION OF ICE CRYSTALS

Before ice can form, water must supercool and ice crystals nucleate. Homogeneous nucleation (without the influence of foreign particles) occurs well below the freezing point, at temperatures that are not observed in water bodies. The temperature of heterogeneous nucleation (nucleation beginning at the surface of foreign particles) depends on the nature of the particles, but it is generally several degrees below the freezing point. Again, super-cooling of this magnitude is not observed in most naturally occurring waters, although some researchers argue that a thin surface layer of water may achieve such supercooling under high rates of heat loss. Nucleation beginning on an ice particle, however, can take place upon only slight supercooling, and it is generally believed that ice particles originating from above the water surface are responsible for the initial onset of ice on the surface of a lake.

Once ice is present, further formation is governed by the rate at which the crystal can grow. This can be very fast. On a cold, still night, when lake water has been cooled to its freezing point and then slightly supercooled on the surface, it is possible to see ice crystals propagating rapidly across the surface. Typically, this form of initial ice formation is such that the crystal c-axes are vertically oriented—in contrast to the usual horizontal orientation of the c-axis associated with later thickening. Under ideal conditions these first crystals may have dimensions of one metre (about three feet) or more. An ice cover composed of such crystals will appear black and very transparent.

THE EFFECTS OF WIND MIXING

If the lake surface is exposed to wind, the initial ice crystals at the surface will be mixed by the agitating effects of wind on the water near the surface, and a layer of small crystals will be created. This layer will act to reduce the mixing, and a first ice cover will be formed consisting of many small crystals. Whether it is composed of large or small crystals, the ice cover, until it grows thick enough to withstand the effects of later winds, may form and dissipate and re-form repeatedly. On larger lakes where the wind prevents a stable ice cover from initially forming, large floes may be formed, and the ice cover may ultimately stabilize as these floes freeze together, sometimes forming large ridges and piles of ice. Ice ridges generally have an underwater draft several times their height above water. If they are moved about by the wind, they may scour the bottom in shallower regions. In some cases—particularly before a stable ice cover forms—wind mixing may be sufficient to entrain ice particles and supercooled water to considerable depths. Water intakes tens of metres deep have been blocked by ice during such events.

THE RATES OF GROWTH

Once an initial layer of ice has formed at the lake surface, further growth proceeds in proportion to the rate at which energy is transferred from the bottom surface of the ice layer to the air above. Because at standard atmospheric pressure the boundary between water and ice is at 0°C (32 °F), the bottom surface is always at the freezing point. If there is no significant flow of heat to the ice from the water below, as is usually the case, all the heat loss through the ice cover will result in ice growth at the bottom. Heat loss through the ice takes place by conduction; designated ϕ in the figure, it is proportional to the thermal conductivity of the ice (k_i) and the temperature difference between the bottom and the top surface of the ice ($T_m - T_s$), and is inversely proportional to the thickness of the ice (h). Heat loss to the air above (also designated ϕ) occurs by a variety of processes, including radiation and convection, but it may be characterized approximately by a bulk transfer coefficient (H_{ia}) times the difference between the surface temperature of the ice and the air temperature ($T_s - T_a$).

(In practice, the top surface of an ice layer is not at the air temperature but somewhere between the air temperature and the freezing point. The exact figures are rarely available, but fortunately the top surface temperature, T_s, is not needed for analysis.)

Assuming that the heat flow through the ice equals the heat flow from the surface of the ice to the air above, the following formula for the thickening of ice may be fashioned:

$$h = \left[\frac{2k}{\rho_i L}(T_m - T_a)t + \left(\frac{k}{H_{ia}}\right)^2 \right]^{1/2} - \frac{k}{H_{ia}}. \qquad (1)$$

In this formula h is the thickness of the ice, T_a is the air temperature, T_m is the freezing point, k is the thermal conductivity of ice (2.24 watts per metre kelvin), ρi is the density of ice (916 kg per cubic metre [57 pounds per cubic foot]), L is the latent heat of fusion (3.34 × 10⁵ joules per kilogram), and t is the time since initial ice formation. The exact value of the bulk transfer coefficient (H_{ia}) depends on the various components of the energy budget, but it usually falls between 10 and 30 watts per square metre kelvin. Higher values are associated with windy conditions and lower values with still air conditions, but, with other information unavailable, a value of 20 watts per square metre kelvin fits data on ice growth quite well. The formula is particularly useful in predicting growth when the ice cover is thin. The first growth rate of the ice cover is proportional to the time since formation; as the ice thickens, however, the top surface temperature more closely approaches the air temperature, and growth proceeds proportional to the square root of time.

If there is a snow layer on top of the ice, it will offer a resistance to the flow of heat from the bottom of the ice surface to the air above. In this case, the incremental thickening rate (that is, the incremental thickening [dh] in an incremental time period [dt]) may be predicted by the following formula:

$$\frac{dh}{dt} = \frac{1}{\rho L}\left(\frac{T_m - T_a}{h_i/k_i + h_s/k_s + 1/H_{ia}}\right), \qquad (2)$$

where h_i is now the ice thickness with thermal conductivity k_i, and h_s is the snow thickness with thermal conductivity k_s. The thermal conductivity of snow depends on its density. It is greater at higher densities, ranging from about 0.1 to 0.5 watt per metre kelvin at densities of 200 to 500 kg per cubic metre (12 to 31 pounds per cubic foot), respectively.

VARIATIONS IN ICE STRUCTURE

When the weight of a snow cover is sufficient to overcome the buoyancy of the ice supporting it, it is usual for the ice to become submerged and for water to flow through cracks in the ice and saturate the snow, which then freezes. This mode of ice growth is different from that analyzed above, but it is quite common, and the ice so formed is known as snow ice. At typical snow densities, a layer of snow about one-half the thickness of the supporting ice will result in the formation of snow ice layers.

As the ice thickens, there is a tendency for crystals with a horizontal c-axis orientation to wedge out adjacent crystals with a vertical c-axis orientation and so become larger in diameter with depth. The resulting structure is one of adjacent columns of single crystals and is termed columnar ice. When a very thin section of the ice is cut and examined with light through crossed polaroid sheets, the crystal structure is clearly seen.

ICE DECAY

As the temperature of the surrounding air and liquid water increases, water sequestered as solid ice reverts to its liquid state. At the microscopic level, such a reversion occurs by thinning and rotting. Wholesale melting over a lake's surface is influenced by the lake's relationship with its shoreline, warm water from rivers, and the amount of solar radiation it can absorb.

THINNING AND ROTTING

In the spring, when average daily air temperatures rise above the freezing point, ice begins to decay. Two processes are active during this period, a dimensional thinning and a deterioration of the ice crystal grains at their boundaries. Thinning of the ice layer is caused by heat transfer and by melting at the top or bottom surface (or both). Deterioration, sometimes called rotting or candling because of the similarity of deteriorating ice crystals to an assembly of closely packed candles, is caused by the absorption of solar radiation. When energy from the Sun warms the ice, melting begins at the grain boundaries because the melting point there is depressed by the presence of impurities that have been concentrated between crystal grains during the freezing process.

Rotting may begin at the bottom or at the top, depending on the particular thermal conditions, but eventually the ice rots throughout its thickness. This greatly reduces the strength of the ice, so that rotten ice will support only a fraction of the load that solid, unrotted ice will support. Thinning and deterioration may occur simultaneously or independently of each other, so that sometimes ice thins without internal deterioration, and sometimes it deteriorates internally with little or no overall thinning. However, both processes usually occur before the ice cover finally breaks up.

Deteriorating ice has a gray, blotchy appearance and looks rotten. Because rotting takes place only by absorption of solar radiation, it progresses only during daylight hours. In addition, the presence of snow or snow ice, which either reflects most solar radiation or absorbs it rapidly in a thin layer, acts to prevent rotting of the ice below until the snow has been completely melted.

MELTING

Melting of lake ice usually occurs first near the shorelines or near the mouths of streams. At these points of contact with inflowing warm water, the ice melts faster than it does at central lake locations, where most melting is caused by the transfer of heat from the atmosphere. Estimates of the rate at which thinning of the main ice cover occurs are usually based on a temperature index method in which a coefficient is applied to the air temperature above freezing.

Water temperature beneath the ice usually reaches its coldest at the time of freeze-up and then gradually warms throughout the winter. The warming is caused by the absorption of some solar radiation that has penetrated the ice cover, by the release of heat that has been stored in bottom sediments during the previous summer, and by warm water inflows. In deep lakes such warming is slight, while in shallow lakes it may amount to several degrees. After snow on the ice has melted in the spring, more solar radiation penetrates the ice cover, so that significant warming may occur. The mixing of warmed water with deteriorated ice is responsible for the very rapid clearing of lake ice at the end of the melt season. On most lakes, the timing of the final clearing of ice is remarkably uniform from year to year, usually varying by less than a week from the long-term average date of clearing.

THE GEOGRAPHIC DISTRIBUTION OF LAKE ICE

The first appearance of lake ice follows by about one month the date at which the long-term average daily air temperature first falls below freezing. Ice appears first in smaller shallow lakes, often forming and melting several times in response to the diurnal variations in air temperature, and finally forms completely as air temperatures remain below the freezing point. Larger lakes freeze over somewhat later because of the longer time required to cool the water. In North America the Canadian-U.S. border roughly coincides with a first freeze-up date of December 1. North of the border freeze-up occurs earlier, as early as October 1 at Great Bear Lake in Canada's Northwest Territories. To the south the year-to-year patterns of freeze-up are ever more erratic until, at latitudes lower than about 45° N, freeze-up may not occur in some years.

In Europe the freeze-up pattern is similar with respect to air temperatures, but the latitudinal pattern shows more variation because much of western Europe is affected by the warming influence of the Gulf Stream. In Central Asia the latitudinal variation is more regular, with first freeze-up occurring about mid-January at 45° N and about October 1 at 72° N. Exceptions to these patterns occur where there are variations in local climate and elevation.

Because of the time required to melt ice that has thickened over the winter, the clearing of lake ice occurs some time after average daily air temperatures rise above freezing. Typically the lag is on the order of one month at latitude 50° N and about six weeks at 70° N. This pattern results in average clearing dates in mid-April at the U.S.-Canadian border and in June and July in the northern reaches of Canada.

ICE FORMATION IN RIVERS

The formation of ice in rivers is more complex than in lakes, largely because of the effects of water velocity and turbulence. As in lakes, the surface temperature drops in response to cooling by the air above. Unlike lakes, however, the turbulent mixing in rivers causes the entire water depth to cool uniformly even after its temperature has fallen below the temperature of maximum density (4°C, or 39°F). The general pattern is one in which the water temperature fairly closely follows the average daily air temperature but with diurnal variations smaller than the daily excursions of air temperature. Once the water temperature drops to the freezing point and further cooling occurs, the water temperature will actually fall below freezing—a phenomenon known as supercooling. Typically the maximum supercooling that is observed is only a few hundredths of a degree Celsius. At this point the introduction of ice particles from the air causes further nucleation of ice in the flow. This freezing action releases the latent heat of fusion, so that the temperature of the water returns toward the freezing point. Ice production is then in balance with the rate of cooling occurring at the surface.

The particles of ice in the flow are termed frazil ice. Frazil is almost always the first ice formation in rivers. The particles are typically about one mm (0.04 inch) or smaller in size and usually in the shape of thin disks. Frazil appears in several types of initial ice formation: thin, sheetlike formations (at very low current velocities); particles that appear to flocculate into larger masses and exhibit a slushlike appearance on the water surface; irregularly shaped "pans" of frazil masses that, while appearing to be shallow, are actually of some depth; and (at high current velocities) a dispersed mixture or slurry of ice particles in the flow.

The supercooling of river water, while amounting to only a few hundredths of a degree Celsius or even less, provides the context for the particles to stick to one another, since under such conditions ice particles are inherently unstable and actively grow into the supercooled water. When they touch one another or some other surface that is cooled below the freezing point, they adhere by freezing. This behaviour causes serious problems at water intakes, where ice particles may adhere and then build up large accumulations that act to block the intake. In rivers and streams, frazil particles also may adhere to the bottom and successively build up a loose, porous layer known as anchor ice. Conversely, if the water temperature then rises above the freezing point, the particles will become neutral and will not stick to one another, so that the flow will be merely one of solid particles in the flowing water. The slightly above-freezing water may also release the bond between anchor ice and the bottom. It is not unusual for anchor ice to form on the bottom of shallow streams at night, when the cooling is great, only to be released the following day under the warming influence of air temperature and solar radiation.

ACCUMULATING ICE COVER

As stated above, frazil forms into pans on the surface of rivers. Eventually these pans may enlarge and freeze together to form larger floes, or they may gather at the leading edge of an ice cover and form a layer of accumulating ice that progresses upstream. The thickness at which such an accumulation collects and progresses upstream depends on the velocity of the flow (V) and is given implicitly in the formula:in which g is acceleration of gravity, ρ

$$V = \left(1 - \frac{h}{H}\right)\sqrt{2g\left(\frac{\rho - \rho_i}{\rho}\right)h}\,, \qquad (3)$$

and ρ_i are the densities of water and ice, respectively, h is the thickness of the accumulating ice, and H is the depth of flow just upstream of the ice cover. As a practical matter, floes arriving at the upstream edge will submerge and pass on downstream if the mean velocity exceeds about 60 cm (24 inches) per second. At certain thicknesses the ice accumulation may not be able to resist the forces exerted by the water flow and by its own weight acting in the downstream direction, and it will thicken by a shoving process until it attains a thickness sufficient to withstand these forces. During very cold periods, freezing of the top layer will provide additional strength by distributing the forces to the shorelines, so that thinner ice covers actually may be better able to withstand the forces acting on them.

As the ice cover accumulates and progresses upstream, it both adds resistance to the flow and displaces a certain volume of water. These two effects cause the depth of the river to be greater upstream, thus reducing the velocity and enabling further upstream progression to occur where previously the current velocity was too high to allow ice cover formation. This phenomenon is termed staging, by reference to its effect of increasing the water level, or "stage." In the process there is a storage of water in the increased depth of the flow upstream, and this somewhat reduces the delivery of water downstream. The breakup of ice in the spring has the opposite effect—that is, the stored water is released and may contribute to a surge of water downstream.

GROWTH OF FIXED ICE COVER

Once the first ice cover has formed and stabilized, further growth is the same as with lake ice: typically columnar crystals grow into the water below, forming a bottom surface that is very smooth. This thickening may be predicted using equation (1), presented earlier for calculating the thickness of lake ice. An exception to this pattern arises when slightly above-freezing water flows beneath the ice cover. When this occurs, the action of the moving water either causes the undersurface to melt or retards the thickening. Since the rate at which melting occurs is proportional to the velocity times the water temperature, the ice cover over areas of higher velocity may be much thinner than in areas of lower velocity. Unfortunately, areas of thinner ice are often not apparent from above and may be dangerous to those traversing it.

In some rivers the initial formation of fixed ice takes place along the shorelines, with the central regions open to the air. The shore ice then gradually widens from the shoreline, and either the central region forms as described above by accumulation of frazil or the two sides of shore ice join.

ICE BUILDUPS

In larger, deeper rivers, frazil produced in upstream reaches may be carried downstream and be transported beneath the fixed ice cover, where it may deposit and form

Ice forms along the banks of Vermont's Ottauquechee River. Some rivers ice over first at the shoreline, where the water is not as deep or dense. Macduff Everton/The Image Bank/Getty Images

large accumulations that are called hanging dams. Such deposits may be of great depth and may actually block large portions of the river's flow. In smaller, shallower streams, similar ice formations may be combinations of shore ice, anchor ice deposits, small hanging-dam-like accumulations, and (over slower-flowing areas) sheet ice.

Ice in smaller streams shows more variation through the winter, since most of the water comes from groundwater inflows during periods between rain. Groundwater is warm and over time may melt the ice formed during very cold periods. At other times all the water in a small stream freezes, and subsequent inflowing water then flows over the surface and also freezes, forming large buildups of ice. These are known as icings, *aufeis* (German), or *naled*s (Russian). Icings may become so thick that they completely block culverts and, in some cases, overflow onto adjacent roads.

DECAY AND ICE JAMS

In late winter, as air temperatures rise above the freezing point, river ice begins to melt owing to heat transfer from above and to the action of the slightly warm water flowing beneath. As occurs in lake ice, river ice also may deteriorate and rot because of absorption of solar radiation. On the undersurface, the action of the turbulent flowing water causes a melt pattern in the form of a wavy relief, with the waves oriented crosswise to the current direction. Eventually, if the ice cover is not subjected to a suddenly increased flow, it may melt in place with little jamming or significant rise in water level. More likely, however, the ice may be moved and form ice jams.

During the spring in very northern areas, and during periods of midwinter thaw in more temperate areas, additional runoff from snowmelt and rain increases the flow in the river. The increased flow raises the water level and may break ice loose from the banks. It also increases the forces exerted on the ice cover. If these forces exceed the strength of the ice, the cover will move and break up and be transported downstream. At some places the quantity of ice will exceed the transport capacity of the river, and an ice jam will form. The jam may then build to thicknesses great enough to raise the water level and cause flooding. Typically, jams form where the slope of the river changes from steeper to milder or where the moving ice meets an intact ice cover—as in a large pool or at the point of outflow into a lake.

Spring breakup jams are usually more destructive than freeze-up jams because of the larger quantities of ice present. Besides causing sudden flooding, the ice itself may collide with structures and cause damage, even to the point of taking out bridges. Sometimes a jam forms, water builds up above it, and the jam breaks loose and moves downstream only to form again. This process may repeat itself several times. In northerly flowing rivers such behaviour is typical, since the upstream ice is freed first and moves toward colder, more stable ice covers.

RIVER ICE MODIFICATION

There are a variety of means of modifying ice in rivers. Icebreaking vessels are used to clear paths for other vessels and occasionally to assist in relieving jams on large rivers. Icebreakers are used extensively in northern Europe and to some extent on the Great Lakes and St. Lawrence River of North America. Dusting the ice cover with a dark material such as coal dust or sand can increase the absorption of solar radiation and thus create areas of weakness that aid in an orderly breakup. Dusting has limited effectiveness, however, if a later snowfall covers the dust layer. Trenching of the ice cover with a ditching or similar machine has been practiced to create a weak zone in areas that are historically prone to jamming. Once ice jams have formed, they are sometimes blasted with explosives. However, if there is no current to transport the ice away after blasting, such measures are usually of little effect.

Ice-retention structures such as floating ice booms are used to hold ice in place and prevent it from moving downstream, where it might cause problems. There have been some attempts to control water releases from dammed reservoirs so as to induce breakup in an orderly manner, but these measures are limited to a narrow range of conditions. Air bubbler systems and flow developers (submerged motor-driven propellers) are used to melt small portions of the ice cover by taking advantage of any thermal reserve, relative to the freezing point, that may exist in the water. These are usually more successful in lakes or enclosed areas than in rivers, since the water temperature in rivers is rarely much above the freezing point.

Wastewater from the cooling of power plants, both fossil-fueled and nuclear, has sometimes been suggested as a source of energy for melting ice downstream of the release points. This method may be advantageous in small areas, but the power requirements for melting extended reaches of ice are immense. Discharges from smaller sources, such as sewage treatment plants, are generally too small to have more than a very localized effect. On the other hand, the water held in reservoirs is often somewhat warmer than freezing, and it can be released in quantities sufficient to result in extended open water downstream— the precise distance depending on how much surface area is required to cool the water back to the freezing point by heat loss to the cold air above.

An icebreaker clears a path for smaller boats near Finland. Travel by river in cold climates may be possible only when ice jams are broken up or removed in some manner. Priit Vesilind/National Geographic/Getty Images.

THE GEOGRAPHIC DISTRIBUTION OF RIVER ICE

Dates of first freeze-up of rivers follow patterns similar to those of lakes, with a tendency for rivers to freeze over somewhat later than smaller lakes. The many factors that affect the freezing process of rivers make generalizations difficult, however. Slower pool-like reaches may freeze over, while more rapidly flowing reaches may remain open well into the winter. Breakup is even more erratic, particularly in the more temperate zones where midwinter thaws may cause a breakup that is followed by another freeze-up and a later breakup as spring temperatures arrive. As a general rule, rivers break up in response to runoff from snowmelt or rain well before lakes clear of ice—although the first shoreline melting in lakes occurs at about the same times as river breakup. In north-flowing rivers, especially in central Russia and western Canada, breakup occurs first in upstream, southerly reaches and then progresses northward with the movement of the spring thaw.

CHAPTER 4
GLACIERS AND ICE SHEETS

A glacier can be defined as any large mass of perennial ice that originates on land by the recrystallization of snow or other forms of solid precipitation and that shows evidence of past or present flow. Exact limits for the terms *large*, *perennial*, and *flow* cannot be set. Except in size, a small snow patch that persists for more than one season is hydrologically indistinguishable from a true glacier. One international group has recommended that all persisting snow and ice masses larger than 0.1 square km (about 0.04 square mile) be counted as glaciers.

Glaciers are classifiable in three main groups. Those that extend in continuous sheets, moving outward in all directions, are called ice sheets if they are the size of Antarctica or Greenland and ice caps if they are smaller. Glaciers confined within a path that directs the ice movement are called mountain glaciers. Those that spread out on level ground or on the ocean at the foot of glaciated regions are called piedmont glaciers or ice shelves, respectively. Glaciers in the third group are not independent and are treated here in terms of their sources; ice shelves with ice sheets, piedmont glaciers with mountain glaciers. A complex of mountain glaciers burying much of a mountain range is called an ice field.

A most interesting aspect of recent geological time (some 30 million years ago to the present) has been the recurrent expansion and contraction of the world's ice cover. These glacial fluctuations influenced geological, climatological, and biological environments and affected the evolution and development of early humans. Almost all of Canada, the northern third of the United States, much of Europe, all of Scandinavia, and large parts of northern Siberia were engulfed by ice during the major glacial stages. At times during the Pleistocene Epoch (about 2.6 million to 11,700 years ago), glacial ice covered 30 percent of the world's land area. At other times the ice cover may have shrunk to less than its present extent. It may not be improper, then, to state that the world is still in an ice age. Because the term "glacial" generally implies ice-age events or Pleistocene time, in this discussion *glacier* is used as an adjective whenever reference is to ice of the present day.

Glacier ice today stores about three-fourths of all the fresh water in the world. Glacier ice covers about 11 percent of the world's land area and would cause a world sea-level rise of about 90 metres (300 feet) if all existing ice melted. Glaciers occur in all parts of the world and at almost all latitudes. In Ecuador, Kenya, Uganda, and Irian Jaya (New Guinea), glaciers even occur at or near the Equator, albeit at high altitudes.

The cause of the fluctuation of the world's glacier cover is still not completely understood. Periodic changes in the heat received from the Sun, caused by fluctuations in the Earth's orbit, are known to correlate with major fluctuations of ice sheet advance and retreat on long time scales. Large ice sheets themselves, however, contain several "instability mechanisms" that may have contributed to the larger changes in world climate. One of these mechanisms is due to the very high albedo, or reflectivity of dry snow to solar radiation. No other material of widespread distribution on the Earth even approaches the albedo of snow. Thus, as an ice sheet expands it causes an ever larger share of the Sun's radiation to be reflected back into space, less is absorbed on the Earth, and the world's climate becomes cooler. Another instability mechanism is implied by the fact that the thicker and more extensive an ice sheet is, the more snowfall it will receive in the form of orographic precipitation (precipitation resulting from the higher altitude of its surface and attendant lower temperature). A third instability mechanism has been suggested by studies of the West Antarctic Ice Sheet. Portions of an ice sheet called ice streams may periodically move rapidly outward, perhaps because of the buildup of a thick layer of wet, deformable material under the ice. Although the ultimate causes of ice ages are not known with certainty, scientists agree that the world's ice cover and climate are in a state of delicate balance.

Only the largest ice masses directly influence global climate, but all ice sheets and glaciers respond to changes in local climate, particularly changes in air temperature or precipitation. The fluctuations of these glaciers in the past can be inferred by features they have left on the landscape. By studying these features, researchers can infer earlier climatic fluctuations.

THE FORMATION AND CHARACTERISTICS OF GLACIER ICE

Glacier ice is an aggregate of irregularly shaped, interlocking single crystals that range in size from a few millimetres to several tens of centimetres. Many processes are involved in the transformation of snowpacks to glacier ice, and they proceed at a rate that depends on wetness and temperature. Snow crystals in the atmosphere are tiny hexagonal plates, needles, stars, or other intricate shapes. In a deposited snowpack these intricate shapes are usually unstable, and molecules tend to evaporate off the sharp (high curvature) points of crystals and be condensed into hollows in the ice grains. This causes a general rounding of the tiny ice grains so that they fit more closely together. In addition, the wind may break off the points of the intricate crystals and thus pack them more tightly. Thus, the density of the snowpack generally increases with time from an initial low value of 50–250 kg per cubic metre (3–15 pounds per cubic foot). The process of evaporation and condensation may continue: touching grains may develop necks of ice that connect them (sintering) and that grow at the expense of other parts of the ice grain, or individual small grains may rotate to fit more tightly together. These processes proceed more rapidly at temperatures near the melting point and more slowly at colder temperatures, but they all result in a net densification of the snowpack. On the other hand, if a strong temperature gradient is present, water molecules may migrate from grain to grain, producing an array of intricate crystal shapes (known as depth hoar) of lowered density. If liquid water is present, the rate of change is many times more rapid because of the melting of ice from grain extremities with refreezing elsewhere, the compacting force of surface tension, refreezing after pressure melting (regulation), and the freezing of water between grains.

This densification of the snow proceeds more slowly after reaching a density of 500–600 kg per cubic metre (31 to 37 pounds per cubic foot), and many of the processes mentioned above become less and less effective. Recrystallization under stress caused by the weight of the overlying snow becomes predominant, and grains change in size and shape in order to minimize the stress on them. This change usually means that large or favourably oriented grains grow at the expense of others. Stresses due to glacier flow may cause further recrystallization. These processes thus cause an increase in the density of the mass and in the size of the average grain.

When the density of the aggregate reaches about 830 to 840 kg per cubic metre (51.8 to 52.4 pounds per cubic foot), the air spaces between grains are sealed off, and the material becomes impermeable to fluids. The time it takes for pores to be closed off is of critical importance for extracting climate-history information from ice cores. With time and the application of stress, the density rises further by the compression of air bubbles, and at great depths the air is absorbed into the ice crystal lattices, and the ice becomes clear. Only rarely in mountain glaciers does the density exceed 900 kg per cubic metre (56 pounds per cubic foot), but at great depths in ice sheets the density may approach that of pure ice (917 kg per cubic metre [57 pounds per cubic foot] at 0°C [32°F] and atmospheric pressure).

Snow that has survived one melting season is called firn (or névé); its density usually is greater than 500 kg per cubic metre (31 pounds per cubic foot) in temperate regions but can be as low as 300 kg per cubic metre (19 pounds per cubic foot) in polar regions. The permeability change at a density of about 840 kg per cubic metre (52 pounds per cubic foot) marks the transition from firn to glacier ice. The transformation may take only three or four years and less than 10 metres (33 feet) of burial in the warm and wet environment of Washington state in North America, but high on the plateau of Antarctica the same process takes several thousand years and burial to depths of about 150 metres (about 500 feet).

Mass Balance

Glaciers are nourished mainly by snowfall, and they primarily waste away by melting and runoff or by the breaking off of icebergs (calving). In order for a glacier to remain at a constant size, there must be a balance between income (accumulation) and outgo (ablation). If this mass balance is positive (more gain than loss), the glacier will grow; if it is negative, the glacier will shrink.

Accumulation refers to all processes that contribute mass to a glacier. Snowfall is predominant, but additional contributions may be made by hoarfrost (direct condensation of ice from water vapour), rime (freezing of supercooled water droplets on striking a surface), hail, the freezing of rain or meltwater, or avalanching of snow from adjacent slopes. Ablation refers to all processes that remove mass from a glacier. In temperate regions, melting at the surface normally predominates. Melting at the base is usually very slight (1 centimetre [0.4 inch] per year or less). Calving is usually the most important process on large glaciers in polar regions and on some temperate glaciers as well. Evaporation and loss by ice avalanches are important in certain special environments; floating ice may lose mass by melting from below.

Because the processes of accumulation, ablation, and the transformation of snow to ice proceed so differently, depending on temperature and the presence or absence of liquid water, it is customary to classify glaciers in terms of their thermal condition. A polar glacier is defined as one that is below the freezing temperature throughout its mass for the entire year; a subpolar (or polythermal) glacier contains ice below the freezing temperature, except for surface melting in the summer and a basal layer of temperate ice; and a temperate glacier is at the melting temperature throughout its mass, but surface freezing occurs in winter. A polar or subpolar glacier may be frozen to its bed (cold-based), or it may be at the melting temperature at the bed (warm-based).

Another classification distinguishes the surface zones, or facies, on parts of a glacier. In the dry-snow zone no surface melting occurs, even in summer; in the percolation zone some surface melting may occur, but the meltwater refreezes at a shallow depth; in the soaked zone sufficient melting and refreezing take place to raise the whole winter snow layer to the melting temperature, permitting runoff; and in the superimposed-ice zone refrozen meltwater at the base of the snowpack (superimposed ice) forms a continuous layer that is exposed at the surface by the loss of overlying snow. These zones are all parts of the accumulation area, in which the mass balance is always positive. Below the superimposed-ice zone is the ablation zone, in which annual loss exceeds the gain by snowfall. The boundary between the accumulation and ablation zones is called the equilibrium line.

The value of the surface mass balance at any point on a glacier can be measured by means of stakes, snow pits, or cores. These values at points can then be averaged over the whole glacier for a whole year. The result is the net or annual mass balance. A positive value indicates growth, a negative value a decline.

HEAT OR ENERGY BALANCE

The mass balance and the temperature variations of a glacier are determined in part by the heat energy received from or lost to the external environment—an exchange that takes place almost entirely at the upper surface. Heat is received from short-wavelength solar radiation, long-wavelength radiation from clouds or water vapour, turbulent transfer from warm air, conduction upward from warmer lower layers, and the heat released by the condensation of dew or hoarfrost or by the freezing of liquid water. Heat is lost by outgoing long-wavelength radiation; turbulent transfer to colder air; the heat required for the evaporation, sublimation, or melting of ice; and conduction downward to lower layers.

In temperate regions, solar radiation is normally the greatest heat source (although much of the incoming radiation is reflected from a snow surface), and most of the heat loss goes to the melting of ice. It is incorrect to think of snow or ice melt as directly related to air temperature; it is the wind structure, the turbulent eddies near the surface, that determines most of the heat transfer from the atmosphere. In polar regions, heat is gained primarily from incoming solar radiation and lost by outgoing long-wavelength radiation, but heat conduction from lower layers and the turbulent transfer of heat to or from the air also are involved.

GLACIER FLOW

In the accumulation area the mass balance is positive year after year. Here the glacier would become thicker and thicker were it not for the compensating flow of ice away from the area. This flow supplies mass to the ablation zone, compensating for the continual loss of ice there.

Glacier flow is a simple consequence of the weight and creep properties of ice. Subjected to a shear stress over time, ice will undergo creep, or plastic deformation. The rate of plastic deformation under constant shear stress is initially high but tapers off to a steady value. If this steady value, the shear-strain rate, is plotted against the stress for many different values of applied stress, a curved graph will result. The curve illustrates what is known as the flow law or constitutive law of ice: the rate of shear strain is approximately proportional to the cube of the shear stress. Often called the Glen flow law by glaciologists, this constitutive law is the basis for all analyses of the flow of ice sheets and glaciers.

As ice tends to build up in the accumulation area of a glacier, a surface slope toward the ablation zone is developed. This slope and the weight of the ice induce a shear stress throughout the mass. In a case with simple geometry, the shear stress can be given by the following formula:

$$\tau = \rho g h \sin \alpha \qquad (4)$$

where τ is the shear stress, ρ the ice density, h the ice thickness, and α the surface slope. Each element of ice deforms according to the magnitude of the shear stress, as determined by (4), at a rate determined by the Glen flow law, stated on page 81. By adding up, or integrating, the shear deformation of each element throughout the glacier thickness, a velocity profile can be produced. It can be given numerical expression as:

$$u_1 = k_1 \sin^3 \alpha h^4 \qquad (5)$$

where u_I is the surface velocity caused by internal deformation and k_I a constant involving ice properties and geometry. In this simple case, velocity is approximately proportional to the fourth power of the depth (h^4). Therefore, if the thickness of a glacier is only slightly altered by changes in the net mass balance, there will be great changes in the rate of flow.

Glaciers that are at the melting temperature at the base may also slide on the bed. Two mechanisms operate to permit sliding over a rough bed. First, small protuberances on the bed cause stress concentrations in the ice, an increased amount of plastic flow, and ice streams around the protuberances. Second, ice on the upstream side of protuberances is subjected to higher pressure, which lowers the melting temperature and causes some of the ice to melt; on the downstream side the converse is true, and meltwater freezes. This process, termed regelation, is controlled by the rate at which heat can be conducted through the bumps. The first process is most efficient with large knobs, and the second process is most efficient with small bumps. Together these two processes produce bed slip. Water-filled cavities may form in the lee of bedrock knobs, further complicating the process. In addition, studies have shown that sliding varies as the basal water pressure or amount changes. Although the process of glacier sliding over bedrock is understood in a general way, none of several detailed theories has been confirmed by field observation. This problem is largely unsolved.

A formula in common use for calculating the sliding speed is:

$$u_2 = \frac{k_2 \sin^2 \alpha}{(p_i - p_a)} \qquad (6)$$

where u_2 is the sliding speed at the base, pi and pa are the ice pressure and water pressure at the base of the ice, and k_2 is another constant involving a measure of the roughness of the bed. The total flow of a glacier can thus be given by the sum of equations (5) and (6), u_1 and u_2. The total sum would be an approximation, because the formulas ignore longitudinal changes in velocity and thickness and other complicating influences, but it has proved to be useful in analyzing situations ranging from small mountain glaciers to huge ice sheets.

Other studies have suggested that many glaciers and ice sheets do not slide on a rigid bed but "ride" on a deforming layer of water-charged sediment. This phenomenon is difficult to analyze because the sediment layer may thicken or thin, and thus its properties may change, depending on the history of deformation. In fact, the process may lead to an unsteady, almost chaotic, behaviour over time. Some ice streams in West Antarctica seem to have exhibited such unsteady behaviour.

MOUNTAIN GLACIERS AND OTHER SMALLER ICE MASSES

In this discussion, the term "mountain glaciers" includes all perennial ice masses other than the Antarctic and Greenland ice sheets. These ice masses are not necessarily associated with mountains. Sometimes the term "small glaciers" is used, but only in a relative sense; a glacier 10,000 square km (4,000 square miles) in surface area would not be called "small" in many parts of the world.

CLASSIFICATION OF MOUNTAIN GLACIERS

Mountain glaciers are generally confined to a more or less marked path directing their movement. The shape of the channel and the degree to which the glacier fills it determine the type of glacier. Valley glaciers are a classic type; they flow at least in part down a valley and are longer than they are wide. Cirque glaciers, short and wide, are confined to cirques, or amphitheatres, cut in the mountain landscape. Other types include transection glaciers or ice fields, which fill systems of valleys, and glaciers in special situations, such as summit glaciers, hanging glaciers, ice aprons, crater glaciers, and regenerated or reconstituted glaciers. Glaciers that spread out at the foot of mountain ranges are called piedmont glaciers. Outlet glaciers are valley glaciers that originate in ice sheets, ice caps, and ice fields. Because of the complex shapes of

Medial moraine of Gornergletscher (Gorner Glacier) in the Pennine Alps near Zermatt, Switz. Jerome Wyckoff

mountain landscapes and the resulting variety of situations in which glaciers can develop, it is difficult to draw clear distinctions among the various types of glaciers.

Mountain glaciers also are classified as polar, subpolar, or temperate and their surfaces by the occurrence of dry-snow, percolation, saturation, and superimposed-ice zones, as for ice sheets.

Surface Features

The snow surface of the accumulation area of a mountain glacier displays the same snow dune and sastrugi features found on ice sheets, especially in winter, but normally these features are neither as large nor as well developed. Where appreciable melting of the snow occurs, several additional features may be produced. During periods of clear, sunny weather, sun cups (cup-shaped hollows usually between 5 and 50 cm [2 and 20 inches] in depth) may develop. On very high-altitude, low-latitude snow and firn fields these may grow into spectacular narrow blades of ice, up to several metres high, called nieves penitentes. Rain falling on the snow surface (or very high rates of melt) may cause a network of meltwater runnels (shallow grooves trending downslope) to develop.

Other features are characteristic of the ablation zone. Below icefalls (steep reaches of a valley glacier), several types of curved bands can be seen. The surface of the glacier may rise and fall in a periodic manner, with the spacing between wave crests approximately equal to the amount of ice flow in a year. Called wave ogives (pointed arches), these arcs result from the great stretching of the ice in the rapidly flowing icefall. The ice that moves through the icefall in summer has more of its surface exposed to melting and is greatly reduced in volume compared with the ice moving through in winter. Dirtband ogives also may occur below icefalls; these are caused by seasonal differences in the amount of dust or by snow trapped in the icefall. Looking down from above, the ogives are invariably distorted into arcs or curves convex downglacier; hence the name ogive.

The ice of the ablation zone normally shows a distinctive layered structure. This can be relict stratification developed by the alternation of dense and light or of clean and dirty snow accumulations from higher on the glacier. This stratification is later subdued by recrystallization accompanying plastic flow. A new layering called foliation is developed by the flow. Foliation is expressed by alternating layers of clear and bubbly or coarse-grained and fine-grained ice. Although the origin of this structure is not fully understood, it is analogous to the process that produces foliated structures in metamorphic rocks.

The ice crystals in strongly deformed, foliated ice invariably have a preferred orientation, relative to the stress directions. In some situations, more often in polar than in temperate ice, the hexagonal axes are aligned perpendicularly to the plane of foliation. This alignment places the crystal glide planes parallel to the planes of (presumed) greatest shearing. In many other locations the hexagonal crystal axes are preferentially aligned in four different directions, none perpendicular to the foliation. This enigmatic pattern has resisted explanation so far.

Crevasses are common to both the accumulation and ablation zones of mountain glaciers, as well as of ice sheets. Transverse crevasses, perpendicular to the flow direction along the centre line of valley glaciers, are caused by extending flow. Splaying crevasses, parallel to the flow in midchannel, are caused by a transverse expansion of the flow. The drag of the valley walls produces marginal crevasses, which intersect the margin at 45°. Transverse and splaying crevasses curve around to become marginal crevasses near the edge of a valley glacier. Splaying and transverse crevasses may occur together, chopping the glacier surface into discrete blocks or towers, called seracs.

Crevasses deepen until the rate of surface stretching is counterbalanced by the rate of plastic flow tending to close the crevasses at depth. Thus, crevasse depths are a function of the rate of stretching and the temperature of the ice. Crevasses deeper than 50 metres (160 feet) are rare in temperate mountains, but crevasses to 100 metres (330 feet) or more in depth may occur in polar regions. Often the crevasses are concealed by a snow bridge, built by accumulations of windblown snow.

MASS BALANCE OF MOUNTAIN GLACIERS

The rate of accumulation and ablation on mountain glaciers depends on latitude, altitude, and distance downwind from sources of abundant moisture, such as the oceans. The glaciers along the coasts of Washington, British Columbia, southeastern Alaska, South Island of New Zealand, Iceland, and southwestern Norway receive prodigious snowfall. Snow accumulation of 3 to 5 metres (10 to 16 feet) of water equivalent in a single season is not uncommon. With this large income, glaciers can exist at low altitudes in spite of very high melt rates. The rate of snowfall increases with increasing altitude; thus, the gradient of net mass balance with altitude is steep. This gradient also expresses the rate of transfer of mass by glacier flow from high to low altitudes and is called the activity index.

Typical of the temperate, maritime glaciers is South Cascade Glacier, in western Washington. Its activity index is high, normally about 17 mm per metre (0.2 inch per foot); the yearly snow accumulation averages about 3.1 metres (about 10 feet) of water-equivalent; and the equilibrium line is at the relatively low altitude of 1,900 metres (6,200 feet). This glacier contains only ablation and saturation zones; the winter chill is so slight that no superimposed ice is formed.

In the maritime environment of southeastern Alaska are many very large glaciers; Bering and Seward-Malaspina glaciers (piedmont glaciers) cover about 5,800 and 5,200 square km (2,200 and 2,000 square miles) in area, respectively. Equilibrium lines are lower than those in Washington State, but the rates of accumulation and ablation and the activity indices are about the same. Because these mountains are high, and some glaciers extend over a great range of altitude, all surface zones except the dry-snow zone are represented.

In more continental (inland) environments, the rate of snowfall is much less, and the summer climate is generally warmer. Thus, glaciers can exist only at high altitudes. High winds may concentrate the meagre snowfall in deep, protected basins, however, allowing glaciers to form even in areas of low precipitation and high melt rates. Glaciers formed almost entirely of drift snow occur at high altitudes in Colorado and in the polar Ural Mountains and are often referred to as Ural-type glaciers. Superimposed ice and soaked zones are found in the accumulation area; in higher areas the percolation zone is found, and in some local extreme areas the dry-snow zone occurs. Because of the decrease in melt rates, continental glaciers in high latitudes occur at lower altitudes and have lower accumulation totals and activity indices. McCall Glacier, in the northwestern part of the Brooks Range in Alaska, has the lowest activity index (two mm per metre) measured in western North America. Glaciers in intermediate climates have intermediate equilibrium-line altitudes, accumulation or ablation totals, and activity indices.

The Flow of Mountain Glaciers

Ice flow in valley glaciers has been studied extensively. The first measurements date from the mid-18th century, and the first theoretical analyses date from the middle of the 19th century. These glaciers generally flow at rates of 0.1 to 2 metres (0.3 to 6.5 feet) per day, faster at the surface than at depth, faster in midchannel than along the margins, and usually fastest at or just below the equilibrium line. Cold, polar glaciers flow relatively slowly, because the constitutive law of ice is sensitive to temperature and because they generally are frozen to their beds. In some high-latitude areas, such as the Svalbard archipelago north of Norway, polythermal glaciers are common; these consist of subfreezing ice overlying temperate ice, and, because they are warm-based, they actively slide on their beds.

The fastest glaciers (other than those in the act of surging) are thick, temperate glaciers in which high subglacial water pressures produce high rates of sliding. Normal temperate glaciers ending on land generally have subglacial water pressures in the range of 50 to 80 percent of the ice pressure, but glaciers that end in the sea may have subglacial water pressures almost equal to the ice pressure—that is, they almost float. The lower reach of Columbia Glacier in southern Alaska, for instance, flows between 20 and 30 metres (66 and 100 feet) per day, almost entirely by sliding. Such a high sliding rate occurs because the glacier, by terminating in the ocean, must have a subglacial water pressure high enough to drive water out of the glacier against the pressure of the ocean water.

Crevasse

Crevasse in the Mozama Glacier on Mount Baker, Washington. Bob and Ira Spring

A crevasse is a fissure or crack in a glacier resulting from stress produced by movement. Crevasses range up to 20 metres (65 feet) wide, 45 metres (148 feet) deep, and several hundred metres long. Most are named according to their positions with respect to the long axis of the glacier. Thus, there are longitudinal crevasses, which develop in areas of compressive stress; transverse crevasses, which develop in areas of tensile stress and are generally curved downstream; marginal crevasses, which develop when the central area of the glacier moves considerably faster than the outer edges; and bergschrund crevasses, which form between the cirque and glacier head. At the terminus of the glacier many crevasses may intersect each other, forming jagged pinnacles of ice called seracs. Crevasses may be bridged by snow and become hidden, and they may close up when the glacier moves over an area with less gradient.

Glacier Hydrology

A temperate glacier is essentially a reservoir that gains precipitation in both liquid and solid form, stores a large share of this precipitation, and then releases it with little loss at a later date. The hydrologic characteristics of this reservoir, however, are complex, because its physical attributes change during a year.

In late spring, the glacier is covered by a thick snowpack at the melting temperature. Meltwater and liquid precipitation must travel through the snowpack by slow percolation until reaching well-defined meltwater channels in the solid ice below. In summer the snowpack becomes thinner and drainage paths within the snow are more defined, so that meltwater and liquid precipitation are transmitted through the glacier rapidly. In winter, snow accumulates, and the surface layer freezes, stopping the movement of meltwater and precipitation at the surface. The rest of the ice reservoir may continue to drain, but in the process the conduits within and under the ice tend to close.

The runoff from a typical Northern Hemisphere temperate glacier reaches a peak in late July or early August. Solar radiation, the chief source of heat to promote melt, reaches a peak in June. The delay in the peak melt rates is primarily because of the changing albedo (surface reflectivity) during the summer. Initially the snow is very reflective and covers the whole glacier, but as the summer wears on the snow becomes wet (less reflective). In addition, more and more ice of much lower albedo is exposed. Thus, even though the incoming radiation decreases during midsummer, the proportion of it that is absorbed to cause melt is greatly increased. Other heat-exchange processes, such as turbulent transfer from warm air, also become more important during midsummer and late summer.

This albedo variation produces a runoff "buffering effect" against unusually wet or dry years. An unusually heavy winter snowpack causes high-albedo snow to persist longer over the glacier in summer; thus, less meltwater is produced. Conversely, an unusually light winter snowfall causes older firn and ice of lower albedo to be exposed earlier in the summer, producing increased melt and runoff. Thus, glaciers naturally regulate the runoff, seasonally and from year to year. When glacier runoff is combined with nonglacier runoff in roughly equal amounts, the result is very stable and even streamflow. This condition is part of the basis for the extensive hydrologic development that is found in regions such as the Alps, Norway, and western Washington State.

Glacier streams are characterized by high sediment concentrations. The sediment ranges from boulders to a distinctive fine-grained material called rock flour, or glacier flour, which is colloidal in size (often less than one micrometre in diameter). The suspended sediment concentration decreases with distance from the glacier, but the rock-flour component may persist for great distances and remain suspended in lakes for many years; it is responsible for the green colour of Alpine lakes. Glacier streams vary in discharge with the time of day, and this variation causes a continual readjustment of the stream channel and the transportation of reworked debris, adding to the sediment load. Rates of glacier erosion (that is, sediment production) are typically on the order of one mm (0.08 inch) per year, averaged over the glacier area, but they are higher in particularly steep terrain or where the bedrock is especially soft.

GLACIER FLOODS

Glacier outburst floods, or *jökulhlaups*, can be spectacular or even catastrophic. These happen when drainage within a glacier is blocked by internal plastic flow and water is stored in or behind the glacier. The water eventually finds a narrow path to trickle out. This movement will cause the path to be enlarged by melting, causing faster flow, more melting, a larger conduit, and so on until all the water is released quite suddenly.

The word *jökulhlaup* is Icelandic in origin, and Iceland has experienced some of the world's most spectacular outburst floods. The 1922 Grimsvötn outburst released about 7.1 cubic km (1.7 cubic miles) of water in a flood that was estimated to have reached almost 57,000 cubic metres (2,000,000 cubic feet) per second. Outburst floods occur in many glacier-covered mountain ranges. Some break out regularly each year, some at intervals of two or more years, and some are completely irregular and impossible to predict.

GLACIER SURGES

Most glaciers follow a regular and nonspectacular pattern of advance and retreat in response to a varying climate. A very different behaviour pattern has been reported for glaciers in certain, but not all, areas. Such glaciers may, after a period of normal flow, or quiescence, lasting 10 to 100 or more years, suddenly begin to flow very rapidly up to 5 metres (16 feet) per hour. This rapid flow, lasting only a year or two, causes a sudden depletion of the upper part of the glacier, accompanied by a swelling and advance of the lower part, although these usually do not reach positions beyond the limits of previous surges. Advances of several kilometres in as many months have been recorded. Even more interesting is the fact that these glaciers periodically repeat cycles of quiescence and activity, irrespective of climate. These unusual glaciers are called surging glaciers.

Although surging glaciers are not rare in some areas (such as the Alaska Range and St. Elias Mountains), they are totally absent in other areas of similar topography, bedrock, climate, and so forth (such as the western Chugach Mountains and Coast Mountains). Furthermore, glaciers of all shapes and sizes, from tiny cirque glaciers to major portions of a large ice cap, have been known to surge. The flow instability that results in glacier surges is generally caused by an abrupt decoupling of the glacier from its bed. This decoupling is the result of a breakdown in the normal subglacier water flow system, but the exact mechanisms that cause some glaciers to surge are not fully understood.

TIDEWATER GLACIERS

Many glaciers terminate in the ocean with the calving of icebergs. Known as tidewater glaciers, these glaciers are the seaward extensions of ice streams originating in ice fields, ice caps, or ice sheets. Some tidewater glaciers are similar to surging glaciers in that they flow at high speeds—as much as 35 metres (115 feet) per day—but they do so continuously. Tidewater glaciers share another characteristic with surging glaciers in that they may advance and retreat periodically, independent of climatic variation.

The physical mechanisms that control the rate of iceberg calving are not yet well understood. Empirical studies of grounded (not floating) tidewater glaciers in Alaska, Svalbard, and elsewhere suggest that the speed of iceberg calving is roughly proportional to water depth at the terminus. This relation can produce an instability and periodic advance-retreat cycles. For example, a glacier terminating in shallow water at the head of a fjord will have a low calving speed that may be exceeded by the ice flow speed, causing advance of the terminus. At the same time, glacial erosion will cause the deposition of sediment as a moraine shoal at the terminus. With time, the glacier will advance, eroding the shoal on the upstream face and depositing sediment on the downstream face. The shoal,

Rock Glacier

A rock glacier is a tonguelike body of coarse rock fragments, found in high mountains above the timberline, that moves slowly down a valley. The rock material usually has fallen from the valley walls and may contain large boulders: it resembles the material left at the terminus of a true glacier. Interstitial ice usually occurs in the centre of rock glaciers. Where the ice approaches the terminus, it melts and releases the rock material, which then forms a steep talus slope. A rock glacier may be 30 metres (100 feet) deep and nearly 1.5 km (about 1 mile) long.

A rock glacier may have wavelike ridges on its surface that curve convexly downstream; these indicate flowage. Maximum movements observed exceed 150 cm (4.5 feet) per year. The method of movement is thought to be either flowage of the interstitial ice or creeping by frost action.

by reducing the depth of the water at the glacier's terminus and thereby inhibiting iceberg calving, will allow the glacier to advance into deep water farther down the fjord. This advance phase is slow—typically 9 to 40 metres (30 to 130 feet) per year—and in an Alaskan fjord it may take a period of 1,000 years or more to cover a typical fjord length of 30 to 130 km (20 to 80 miles).

Such a glacier, in an extended position and terminating in shallow water on a moraine shoal, is in an unstable situation. If, for some reason, the terminus retreats slightly, the deeper water upstream of the shoal will cause an increase in iceberg calving; this will result in further retreat into deeper water, which will further increase the calving until the calving speed becomes so high that the normal processes of glacier flow cannot compensate. A rapid, irreversible retreat will result until the glacier reaches shallow water back at the head of the fjord. In contrast to the slow advance phase, the retreat phase may take only a few decades. The fastest glacier retreats observed during historical time (for instance, the opening of Glacier Bay, Alaska), as well as those inferred during the demise of the great Quaternary ice sheets, were caused by this mechanism. Information on the advance and retreat of tidewater glaciers should not be used to infer climatic change, however.

THE GREAT ICE SHEETS

Two great ice masses, the Antarctic and Greenland ice sheets, stand out in the world today and may be similar in many respects to the large Pleistocene ice sheets. About 99 percent of the world's glacier ice is in these two ice masses — 91 percent in Antarctica alone.

THE ANTARCTIC ICE SHEET

The bedrock of the continent of Antarctica is almost completely buried under ice. Mountain ranges and isolated nunataks (a term derived from Greenland's Inuit language, used for individual mountains surrounded by ice) locally protrude through the ice. Extensive in area are the ice shelves, where the ice sheet extends beyond the land margin and spreads out to sea. The ice sheet, with its associated ice shelves, covers an area of 13,829,000 square km (5,340,000 square miles); exposed rock areas total less than 200,000 square km (77,000 square miles). The mean thickness of the ice is about 1,829 metres (6,000 feet) and the volume of ice more than 25.4 million cubic km (6 million cubic miles).

The land surface beneath the ice is below sea level in many places, but this surface is depressed because of the weight of the ice. If the ice sheet were melted, uplift of the land surface would eventually leave only a few deep troughs and basins below sea level—even though the sea level itself also would rise about 80 metres (260 feet) from the addition of such a large amount of water. Because of the thick ice cover, Antarctica has by far the highest mean altitude of the continents (2 km [1.3 miles]); all other continents have mean altitudes less than 1 km (0.6 mile).

Antarctica can be divided into three main parts: the smallest and the mildest in climate is the Antarctic Peninsula, extending from latitude 63° S off the tip of South America to a juncture with the main body of West Antarctica at a latitude of about 74° S. The ice cover of the Antarctic Peninsula is a complex of ice caps, piedmont and mountain glaciers, and small ice shelves.

The part of the main continent lying south of the Americas, between longitudes 45° W and 165° E, is characterized by irregular bedrock and ice-surface topography and numerous nunataks and deep troughs. Two large ice shelves occur in West Antarctica: the Filchner-Ronne Ice Shelf (often considered to be two separate ice shelves), south of the Weddell Sea, and the Ross Ice Shelf, south of the Ross Sea. Each has an area exceeding 500,000 square km (193,000 square miles).

The huge ice mass of East Antarctica, about 10,200,000 square km (about 4,000,000 square miles), is separated from West Antarctica by the Transantarctic Mountains. This major mountain range extends from the eastern margin of the Ross Ice Shelf almost to the Ronne-Filchner Ice Shelf. The bedrock of East Antarctica is approximately at sea level, but the ice surface locally exceeds 4,000 metres (13,100 feet) above sea level on the highest parts of the polar plateau.

At the South Pole the snow surface is 2,800 metres (9,200 feet) in altitude, and the mean annual temperature is about -50°C (-58°F), but at the Russian Vostok Station (78°27′ S, 106°52′ E), 3,500 metres (about 11,500 feet) above sea level, the mean annual temperature is -58°C (-73°F), and in August 1960 (the winter season) the temperature reached a low of -88.3°C (-127°F). The temperatures on the polar plateau of East Antarctica are by far the coldest on Earth; the climate of the Arctic is quite mild by comparison. Along the coast of East or West Antarctica, where the climate is milder, mean annual temperatures range from -20 to -9°C (-4 to 16°F), but temperatures exceed the melting point only for brief periods in summer, and then only slightly. Katabatic (drainage) winds, however, are very strong along the coast; the mean annual wind speed at Commonwealth Bay is 20 metres per second (45 miles per hour).

THE GREENLAND ICE SHEET

The Greenland Ice Sheet, though subcontinental in size, is huge compared with other glaciers in the world except that of Antarctica. Greenland is mostly covered by this single large ice sheet (1,730,000 square km [668,000 square miles]), while isolated glaciers and small ice caps totaling between 76,000 and 100,000 square km (29,000 and 39,000 square miles) occur around the periphery. The ice sheet is almost 2,400 km (about 1,500 miles) long in a north-south direction, and its greatest width is 1,100 km (680 miles) at a latitude of 77° N, near its northern margin. The mean altitude of the ice surface is 2,135 metres (7,000 feet). The term Inland Ice, or, in Danish, *Indlandsis*, is often used for this ice sheet.

The bedrock surface is near sea level over most of the interior of Greenland, but mountains occur around the periphery. Thus, this ice sheet, in contrast to the Antarctic Ice Sheet, is confined along most of its margin. The ice

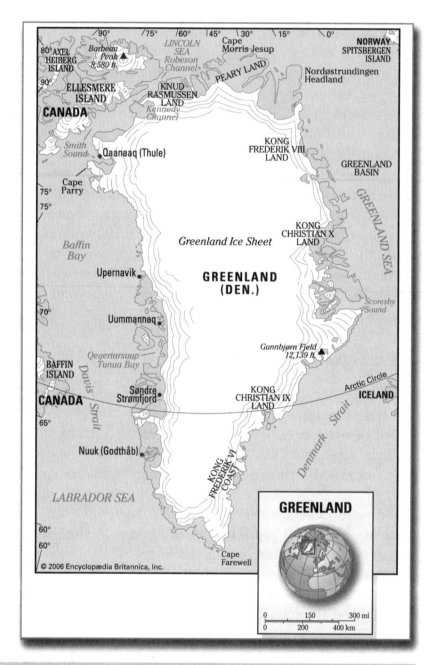

Map of Greenland highlighting the major geographic regions and the locations of human settlement.

surface reaches its greatest altitude on two north-south elongated domes, or ridges. The southern dome reaches almost 3,000 metres (about 9,800 feet) at latitudes 63°–65° N; the northern dome reaches about 3,290 metres (10,800 feet) at about latitude 72° N. The crests of both domes are displaced east of the centre line of Greenland.

The unconfined ice sheet does not reach the sea along a broad front anywhere in Greenland, and no large ice shelves occur. The ice margin just reaches the sea, however, in a region of irregular topography in the area of Melville Bay southeast of Thule. Large outlet glaciers, which are restricted tongues of the ice sheet, move through bordering valleys around the periphery of Greenland to calve off into the ocean, producing the numerous icebergs that sometimes penetrate North Atlantic shipping lanes. The best known of these is the Jakobshavn Glacier, which, at its terminus, flows at speeds of 20 to 22 metres (66 to 72 feet) per day.

The climate of the Greenland Ice Sheet, though cold, is not as extreme as that of central Antarctica. The lowest mean annual temperatures, about -31°C (-24°F), occur on the north-central part of the north dome, and temperatures at the crest of the south dome are about -20°C (-4°F).

The ice sheet is the largest and possibly the only relict of the Pleistocene glaciations in the Northern Hemisphere. In volume it contains 12 percent of the world's glacier ice, and, if it melted, sea level would rise 20 feet (6 metres). In the 1970s and early 1980s the Greenland Ice Sheet Program was organized by scientists from the United States, Denmark, and Switzerland. Deep ice cores from the Greenland Ice Sheet were obtained for comparison with deep cores from the Antarctic ice mass to gain a better understanding of the factors controlling present and past ice mass dynamics, atmospheric processes, and the response of ice sheets to climatic change and to determine whether the past changes in climate were global or regional in character.

ACCUMULATION AND ABLATION OF THE ICE SHEETS

The size and thickness of an ice sheet depends on the amount of precipitation it receives and the amount of material it loses to melting, calving, and other processes. An ice sheet grows in years where accumulation exceeds ablation and declines in years where ablation exceeds accumulation.

The rate of precipitation on the Antarctic Ice Sheet is so low that it may be called a cold desert. Snow accumulation over much of the vast polar plateau is less than five cm (two inches) water equivalent per year. Only around the margin of the continent, where cyclonic storms penetrate frequently, does the accumulation rise to values of more than 30 cm (12 inches). The mean for Antarctica is 15 cm (6 inches) or less. In Greenland values are higher: less than 15 cm in a comparatively small area of north-central Greenland, 30 cm along the crests of the domes, and more than 80 cm (32 inches) along the southeast and southwest margins; the mean annual snow accumulation is about 37 cm (15 inches) of water equivalent.

Snow accumulation occurs mainly as direct snowfall when cyclonic storms move inland. At high altitudes on the Greenland Ice Sheet and in central Antarctica, ice crystals form in the cold air during clear periods and slowly settle out as fine "diamond dust." Hoarfrost and rime deposition are generally minor items in the snow-accumulation totals. It is almost impossible to measure the precipitation directly in these climates; precipitation gauges are almost useless for the measurement of blowing snow, and the snow is blown about almost constantly in some areas. The thickness and density of snow deposited on the ground equals precipitation plus hoarfrost and rime deposition, less evaporation, less snow blown away, and plus snow blown in from somewhere else. The last two phenomena are thought to cancel each other approximately—except in the coastal areas, where fierce drainage, or katabatic, winds move appreciable quantities of snow out to sea.

The snow surface may be smooth where soft powder snow is deposited with little wind, or very hard packed and rough when high winds occur during or after snowfall. Two features are prominent: snow dunes are depositional features resembling sand dunes in their several shapes; sastrugi are jagged erosional features (often cut into snow dunes) caused by strong prevailing winds that occur after snowfall. Sharp, rugged sastrugi, which can be one to two metres (6.5 feet) high, make travel by vehicle or on foot difficult. The annual snow layers exposed in the side of a snow pit can usually be distinguished by a low density layer (depth hoar) that forms by the burial of surface hoarfrost or by metamorphism of the snow deposited in the fall at a time when the temperature is changing rapidly.

Almost all of the Antarctic Ice Sheet lies within the dry-snow zone. The percolation, soaked, and superimposed ice zones occur only in a very narrow strip in a small area along the coast. In Greenland only the central part of the northern half of the ice sheet, or about 30 percent of the total area, is within the dry-snow zone. Almost half of the area of the Greenland ice sheet is considered to be in the percolation zone. In flat areas near the equilibrium line, especially in west-central Greenland, there are notorious snow swamps, or slush fields, in summer; some of this water runs off, but much of it refreezes.

The ice sheets lose material by several processes, including surface melting, evaporation, wind erosion (deflation), iceberg calving, and the melting of the bottom surfaces of floating ice shelves by warmer seawater.

In Antarctica, calving of ice shelves and outlet glacier tongues clearly predominates among all the processes of ice loss, but calving is very episodic and cannot be measured accurately. The amount of surface melt and evaporation is small, amounting to about 22 cm (about 9 inches) of ice lost from a 5 km (3 mile) ring around half the continent. Wind erosion is difficult to evaluate but probably accounts for only a very small loss in the mass balance. The undersides of ice shelves near their outer margins are subject to melting by the ocean water. The rate of melting decreases inland, and at that point some freezing of seawater onto the base of the ice shelves must occur, but farther inland, near the grounding line, the tidal circulation of warm seawater may produce basal melting.

In Greenland, surface melt is more important, calving is less so, and undershelf melting is important only on floating glacier tongues (seaward projections of a glacier). Most of the calving is from the termini of a relatively few large, fast-moving outlet glaciers. In Greenland, vertical-walled melt pits in the ice are a well-known feature of the ice surface at the ablation zone. Ranging from a few millimetres to a metre (3 feet) in diameter, these pits are floored with a dark, silty material called cryoconite, once thought to be of cosmic origin but now known to be largely terrestrial dust. The vertical melting of the holes is due to the absorption of solar radiation by the dark silt, possibly augmented by biological activity.

NET MASS BALANCE

Because two great ice sheets contain 99 percent of the world's ice, it is important to know whether this ice is growing or shrinking under present climatic conditions. Although just such a determination was a major objective of the International Geophysical Year (1957–58) and more has been learned each year since, even the sign of the net mass balance has not yet been determined conclusively.

It appears that accumulation on the surface of the Antarctic Ice Sheet is approximately balanced by iceberg calving and basal melting from the ice shelves. Compilations from many authors and the Intergovernmental Panel on Climate Change (IPCC), *Third Scientific Assessment* (2001), suggest the following average values, given in gigatons (billions of tons) per year (1 gigaton is equivalent to 1.1 cubic km [0.26 cubic mile] of water):

Accumulation
Accumulation on grounded ice + 1,829 ± 87
Accumulation on grounded ice
and ice shelves + 2,233 ± 86

Ablation
Calving of ice shelves and glaciers - 2,072 ± 304
Bottom melting, ice shelves - 540 ± 218
Melting and runoff - 10 ± 10
Net mass balance - 389 ± 384

The net difference, however, is on the same order as the margin of error in estimating the various quantities. Furthermore, some authors have suggested that the values stated above for calving and ice-shelf melting are too high and that the discharge of ice to the sea, as measured by ice-flow studies, is clearly less than the accumulation. Thus, even the sign of the net balance is not well defined. It appears that the net balance of the grounded portion of the Antarctic Ice Sheet is positive, while that of the floating ice shelves is negative. Studies of fluctuations in the extent of floating ice have been inconclusive.

The net mass balance of the Greenland Ice Sheet also appears to be close to zero, but here, too, the margin of error is too large for definite conclusions. The estimated balance is as follows, again from the IPCC and in gigatons per year.

Accumulation
Snow accumulation 522 ± 21

Ablation
Iceberg calving - 235 ± 33
Melting and runoff - 297 ± 32
Bottom melting - 32 ± 3
Net mass balance - 42 ± 51

Uncertainties in the quantities given above are due to the difficulty of analyzing the spatial and temporal distributions of accumulation, the relatively few annual measurements of iceberg calving, and a lack of knowledge of the amount of surface meltwater that refreezes in the cold snow and ice at depth. Many of the outlet glaciers and portions of the ice-sheet margin in the southwestern part of Greenland, where many observations have been made, have stopped the retreats that were observed from the 1950s through the 1970s. After a period of relative stability and advance during the 1980s, glacier retreats have both resumed and accelerated in Greenland since the mid-1990s.

THE FLOW OF THE ICE SHEETS

In general, the flow of the Antarctic and Greenland ice sheets is not directed radially outward to the sea. Instead, ice from central high points tends to converge into discrete drainage basins and then concentrate into rapidly flowing ice streams. (Such so-called streams are currents of ice that move several times faster than the ice on either side of them.) The ice of much of East Antarctica has a rather simple shape with several subtle high points or domes. Greenland resembles an elongated dome, or ridge, with two summits. West Antarctica is a complex of converging and diverging flow because of the jumble of ridges and troughs in the subglacial bedrock and the convergence of ice streams.

Flow rates in the interior of an ice sheet are very low, being measured in centimetres or metres per year, because the surface slope is minuscule and the ice is very cold. As the ice moves outward, the rate of flow increases to a few tens of metres per year, and this rate of flow increases still further, up to 1 km (0.6 mile) per year, as the flow is channeled into outlet glaciers or ice streams. Ice shelves continue the flow and even cause it to increase, because ice spreads out in ever thinner layers. At the edge of the Ross Ice Shelf, ice is moving out about 900 metres (2,950 feet) per year toward the ocean.

This simple picture of ice flow is made more complicated by the dependence of the flow law of ice on temperature. Because a temperature increase of about 15°C (27°F) causes a 10-fold increase in the deformation rate of ice, the temperature distribution of an ice sheet partly determines its flow structure. The cold ice of the central part of an ice sheet is carried down into warmer zones. This shift modifies the static temperature distribution, and the shear deformation is concentrated in a thin zone of warmer ice at the base. The forward velocity may be almost uniform throughout the depth to within a few tens or hundreds of metres from the bedrock.

Another important effect on ice flow is the heat produced by friction, caused by the sliding of the ice on bedrock or by internal shearing within the basal ice. If a portion of the ice sheet deforms more rapidly than its surroundings, the slight amount of extra heat production raises the temperature of this portion, causing it to deform even more readily. This increased deformation may explain the phenomena of ice streams, which are very effective in moving ice from large drainage areas of Antarctica and Greenland out to ice shelves or to the sea. It is known that at least one Antarctic ice stream moves rapidly on a layer of water-charged deforming sediment; a nearby ice stream appears to have ceased rapid movement in the past several hundred years, perhaps owing to loss of its sediment layer.

Ross Ice Shelf

The world's largest body of floating ice, the Ross Ice Shelf, lies at the head of Ross Sea, itself an enormous indentation in the continent of Antarctica. The ice shelf lies between about 155° W and 160° E longitude and about 78° S and 86° S latitude. The current estimate of its area is about 472,000 square km (182,000 square miles), making it roughly the size of the Yukon territory in Canada. The shelf has served as an important gateway for explorations of the Antarctic interior, including those carried out by many of the most famous expeditions.

The great white barrier wall of the shelf's front, first seen in 1841 by the British polar explorer James Clark Ross, rises in places to 50 or 60 metres (160 or 200 feet) high and stretches about 800 km (500 miles) between fixed "anchor points" on Ross Island to the west and the jutting Edward VII Peninsula on the east. With its immense, gently undulating surface reaching back nearly 950 km (600 miles) southward into the heart of Antarctica, the Ross Ice Shelf provides the best surface approach into the continental interior. The McMurdo Sound region on the shelf's western edge thus became the headquarters for Robert F. Scott's 1911–12 epic sledging trip to the South Pole and also served several Antarctic research programs later in the century. The eastern barrier regions of the ice shelf were headquarters for the Norwegian Roald Amundsen's first attainment of the South Pole on Dec. 14, 1911; for Richard E. Byrd's three U.S. expeditions of 1928–41 at Little America I–III stations; and for several subsequent expeditions and research programs.

The Ross Ice Shelf is fed primarily by giant glaciers, or ice streams, that transport ice down to it from the high polar ice sheet of East and West Antarctica. The ice shelf has been likened to a vast triangular raft because it is relatively thin and flexible and is only loosely attached to adjoining lands. Giant rifts develop behind the ice shelf's barrier wall and occasionally rupture completely to spawn the huge tabular icebergs that are so characteristic of the Antarctic Ocean. Thus, although the barrier's position appears almost stationary, it actually undergoes continual change by calving and melting that accompany northward movement of the ice shelf.

The shelf's mean ice thickness is about 330 metres (1,100 feet) along a line at about 79° S latitude. In a southward direction along about 168° W longitude, the ice shelf's thickness gradually increases to more than 700 metres (2,300 feet). Estimates suggest that at distances of 160 to 320 km (100 to 200 miles) inland from the barrier, 380 to 500 mm (15 to 20 inches) of ice may be added to the shelf each year by bottom freezing. Melting on the bottom of the ice shelf also occurs, and oceanographic data suggest that the net effect is the dissolution of the ice shelf by about 120–220 cm (47–87 inches) per year.

THE INFORMATION FROM DEEP CORES

Most of the Antarctic and Greenland ice sheets are below freezing throughout. Continuous cores, taken in some cases to the bedrock below, allow the sampling of an ice sheet through its entire history of accumulation. Records obtained from these cores represent exciting new developments in paleoclimatology and paleoenvironmental studies. Because there is no melting, the layered structure of the ice preserves a continuous record of snow accumulation and chemistry, air temperature and chemistry, and fallout from volcanic, terrestrial, marine, cosmic, and

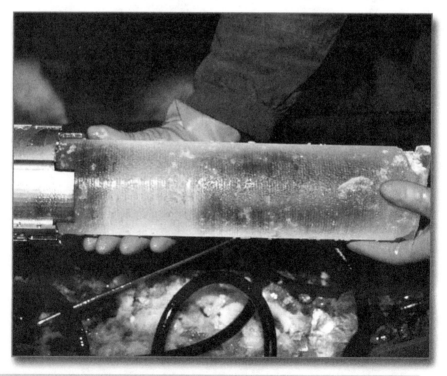

A glaciologist in Antarctica examines an ice core dating to 1840. Testing performed on the layers of an ice core reveal much about the climate and atmosphere of the past. Vin Morgan/AFP/Getty Images.

man-made sources. Actual samples of ancient atmospheres are trapped in air bubbles within the ice. This record extends back more than 400,000 years.

Near the surface it is possible to pick out annual layers by visual inspection. In some locations, such as the Greenland Ice-core Project/Greenland Ice Sheet Project 2 (GRIP/GISP2) sites at the summit of Greenland, these annual layers can be traced back more than 40,000 years, much like counting tree rings. The result is a remarkably high-resolution record of climatic change. When individual layers are not readily visible, seasonal changes in dust, marine salts, and isotopes can be used to infer annual chronologies. Precise dating of recent layers can be accomplished by locating radioactive fallout from known nuclear detonations or traces of volcanic eruptions of known date. Other techniques must be used to reconstruct a chronology from some very deep cores. One method involves a theoretical analysis of the flow. If the vertical profile of ice flow is known, and if it can be assumed that the rate of accumulation has been approximately constant through time, then an expression for the age of the ice as a function of depth can be developed.

A very useful technique for tracing past temperatures involves the measurement of oxygen isotopes—namely, the ratio of oxygen-18 to oxygen-16. Oxygen-16 is the dominant isotope, making up more than 99 percent of all natural oxygen; oxygen-18 makes up 0.2 percent. However, the exact concentration of oxygen-18 in precipitation, particularly at high latitudes, depends on the temperature. Winter snow has a smaller oxygen-18–oxygen-16 ratio than does summer snow. A similar isotopic method for inferring precipitation temperature is based on measuring the ratio of deuterium (hydrogen-2) to normal hydrogen (hydrogen-1). The relation between these oxygen and hydrogen isotopic ratios, termed the deuterium excess, is useful for inferring conditions at the time of evaporation and precipitation. The temperature scale derived from isotopic measurements can be calibrated by the observable temperature-depth record near the surface of ice sheets.

Results of ice core measurements are greatly extending the knowledge of past climates. For instance, air samples taken from ice cores show an increase in methane, carbon dioxide, and other "greenhouse gas" concentrations with the rise of industrialization and human population. On a longer time scale, the concentration of carbon dioxide in the atmosphere can be shown to be related to atmospheric temperature (as indicated by oxygen and hydrogen isotopes)—thus confirming the global-warming greenhouse effect, by which heat in the form of long-wave infrared radiation is trapped by atmospheric carbon dioxide and reflected back to the Earth's surface.

Perhaps most exciting are recent ice core results that show surprisingly rapid fluctuations in climate, especially during the last glacial period (160,000 to 10,000 years ago) and probably in the interglacial period that preceded it. Detectable variations in the dustiness of the atmosphere (a function of wind and atmospheric circulation), temperature, precipitation amounts, and other variables show that, during this time period, the climate frequently alternated between full-glacial and nonglacial conditions in less than a decade. Some of these changes seem to have occurred as sudden climate fluctuations, called Dansgaard-Oeschger events, in which the temperature jumped 5 to 7°C (9 to 13°F), remained in that state for a few years to centuries, jumped back, and repeated the process several times before settling into the new state for a long time— perhaps 1,000 years. These findings have profound and unsettling implications for the understanding of the coupled ocean-atmosphere climate system.

THE RESPONSE OF GLACIERS TO CLIMATIC CHANGE

The relationship of glaciers and ice sheets to fluctuations in climate is sequential. The general climatic or meteorological environment determines the local mass and heat-exchange processes at the glacier surface, and these in turn determine the net mass balance of the glacier. Changes in the net mass balance produce a dynamic response—that is, changes in the rate of ice flow. The dynamic response causes an advance or retreat of the terminus, which may produce lasting evidence of the change in the glacier margin. If the local climate changes toward increased winter snowfall rates, the net mass balance becomes more positive, which is equivalent to an increase in ice thickness. The rate of glacier flow depends on thickness, so that a slight increase in thickness produces a larger increase in ice flow. This local increase in thickness and flow propagates downglacier, taking some finite amount of time. When the change arrives at the terminus, it causes the margin of the glacier to extend farther downstream. The result is known as a glacier fluctuation—in this case an advance—and it incorporates the sum of all the changes that have taken place up-glacier during the time it took them to propagate to the terminus.

The process, however, cannot be traced backward with assurance. A glacier advance can, perhaps, be related to a period of positive mass balances, but to ascertain the meteorological cause is difficult because either increased snowfall or decreased melting can produce a positive mass balance.

The dynamic response of glaciers to changes in mass balance can be calculated several ways. Although the complete, three-dimensional equations for glacier flow are difficult to solve for changes in time, the effect of a small change or perturbation in climate can be analyzed readily. Such an analysis involves the theory of kinematic waves, which are akin to small pulses in one-dimensional flow systems such as floods in rivers or automobiles on a crowded roadway. The length of time it takes the glacier to respond in its full length to a change in the surface mass balance is approximately given as the ratio of ice thickness to (negative) mass balance at the terminus. The time scale for mountain glaciers is typically on the order of 10 to 100 years—although for thick glaciers or those with low ablation rates it can be much longer. Ice sheets normally have time scales several orders of magnitude longer.

GLACIERS AND SEA LEVEL

Sea level is currently rising at about 1.8 mm (0.07 inch) per year. Between 0.3 and 0.7 mm (0.01 to 0.03 inch) per year has been attributed to thermal expansion of ocean water, and most of the remainder is thought to be caused by the melting of glaciers and ice sheets on land. There is concern that the rate in sea-level rise may increase markedly in the future owing to global warming. Unfortunately, the state of the mass balance of the ice on the Earth is poorly known, so the exact contributions of the different ice masses to rising sea level is difficult to analyze. The mountain (small) glaciers of the world are thought to be contributing 0.2 to 0.4 mm (0.01 to 0.02 inch) per year to the rise. Yet the Greenland Ice Sheet is thought to be close to balance, the status of the Antarctic Ice Sheet is uncertain, and, although the floating ice shelves and glaciers may be in a state of negative balance, the melting of floating ice should not cause sea level to rise, and the grounded portions of the ice sheets seem to be growing. Thus, the cause of sea-level rise is still not well understood.

With global warming, the melting of mountain glaciers will certainly increase, although this process is limited: the total volume of small glaciers is equivalent to only about 0.6 metre (2 feet) of sea-level rise. Melting of the marginal areas of the Greenland Ice Sheet will likely occur under global warming conditions, and this will be accompanied by the drawing down of the inland ice and increased calving of icebergs; yet these effects may be counterbalanced to some extent by increased snow precipitation on the inland ice. The Antarctic Ice Sheet, on the other hand, may actually serve as a buffer to rising sea level. Increased melting of the marginal areas will probably be exceeded by increased snow accumulation due to the warmer air (which holds more moisture) and decreased sea ice (bringing moisture closer to the ice sheet). Modeling studies that predict sea-level rise up to the time of the doubling of greenhouse gas concentrations (i.e., concentrations of atmospheric carbon dioxide, methane, nitrous oxide, and certain other gases) about the year 2050 suggest a modest rise of about 0.3 metre (1 foot).

CHAPTER 5

GLACIAL LANDFORMS

G lacial landforms are any product of flowing ice and meltwater. Such landforms are being produced today in glaciated areas, such as Greenland, Antarctica, and many of the world's higher mountain ranges. In addition, large expansions of present-day glaciers have recurred during the course of Earth history. At the maximum of the last ice age, which ended about 20,000 to 15,000 years ago, more than 30 percent of the Earth's land surface was covered by ice. Consequently, if they have not been obliterated by other landscape-modifying processes since that time, glacial landforms may still exist in regions that were once glaciated but are now devoid of glaciers.

Periglacial features, which form independently of glaciers, are nonetheless a product of the same cold climate that favours the development of glaciers, and so are treated here as well.

GENERAL CONSIDERATIONS

In order to gain a solid understanding of the different landforms produced by glaciers and their meltwater, it is helpful to discuss the glacial environment and the processes responsible for the formation of such structures.

GLACIERS AND TOPOGRAPHY

There are numerous types of glaciers, but it is sufficient here to focus on

Eskers, narrow ridges of gravel and sand left by a retreating glacier, wind through western Nunavut, Canada, near the Thelon River. © Richard Alexander Cooke III

two broad classes: mountain, or valley, glaciers and continental glaciers, or ice sheets (including ice caps).

Generally, ice sheets are larger than valley glaciers. The main difference between the two classes, however, is their relationship to the underlying topography. Valley glaciers are rivers of ice usually found in mountainous regions, and their flow patterns are controlled by the high relief in those areas. In map view, many large valley glacier systems, which have numerous tributary glaciers that join to form a large "trunk glacier," resemble the roots of a plant. Pancakelike ice sheets, on the other hand, are continuous over extensive areas and completely bury the underlying landscape beneath hundreds or thousands of metres of ice. Within continental ice sheets, the flow is directed more or less from the centre outward. At the periphery, however, where ice sheets are much thinner, they may be controlled by any substantial relief existing in the area. In this case, their borders may be lobate on a scale of a few kilometres, with tonguelike protrusions called outlet glaciers. Viewed by themselves, these are nearly indistinguishable from the lower reaches of a large valley glacier system. Consequently, many of the landforms produced by valley glaciers and continental ice sheets are similar or virtually identical, though they often differ in magnitude. Nonetheless, each type of glacier produces characteristic features and thus warrants separate discussion.

GLACIAL EROSION

Two processes, internal deformation and basal sliding, are responsible for the movement of glaciers under the influence of gravity. The temperature of glacier ice is a critical condition that affects these processes. For this reason, glaciers are classified into two main types, temperate and polar, according to their temperature regime. Temperate glaciers are also called isothermal glaciers, because they exist at the pressure-melting point (the melting temperature of ice at a given pressure) throughout their mass. The ice in polar, or cold glaciers, in contrast, is below the pressure-melting point. Some glaciers have an intermediate thermal character. For example, subpolar glaciers are temperate in their interior parts, but their margins are cold-based. This classification is a broad generalization, however, because the thermal condition of a glacier may show wide variations in both space and time.

Internal deformation, or strain, in glacier ice is a response to shear stresses arising from the weight of the ice (ice thickness) and the degree of slope of the glacier surface. Internal deformation occurs by movement within and between individual ice crystals (slow creep) and by brittle failure (fracture), which arises when the mass of ice cannot adjust its shape rapidly enough by the creep process to take up the stresses affecting it. The relative importance of these two processes is greatly influenced by the temperature of the ice. Thus, fractures due to brittle failure under tension, known as crevasses, are usually much deeper in polar ice than they are in temperate ice.

The temperature of the basal ice is an important influence upon a glacier's ability to erode its bed. When basal temperatures are below the pressure-melting point, the ability of the ice mass to slide on the bed (basal sliding) is inhibited by the adhesion of the basal ice to the frozen bed beneath. Basal sliding is also diminished by the greater rigidity of polar ice. This reduces the rate of creep, which, in turn, reduces the ability of the more rigid ice to deform around obstacles on the glacier bed. Thus, the flow of cold-based glaciers is predominantly controlled by internal deformation, with proportionately low rates of basal sliding. For this reason, rates of abrasion are commonly low beneath polar glaciers, and slow rates of erosion commonly result. Equally, the volume of meltwater is frequently very low, so that the extent of sediments and landforms derived from polar glaciers is limited.

Temperate glaciers, being at the pressure-meeting point, move by both mechanisms, with basal sliding being the more important. It is this sliding that enables temperate glaciers to erode their beds and carve landforms so effectively. Ice is, however, much softer and has a much lower shear strength than most rocks, and pure ice alone is not capable of substantially eroding anything other than unconsolidated sediments. Most temperate glaciers have a basal debris zone from several centimetres to a few metres thick that contains varying amounts of rock debris in transit. In this respect, glaciers act rather like sheets of sandpaper. While the paper itself is too soft to sand wood, the adherent hard grains make it a powerful abrasive system.

The analogy ends here, however, for the rock debris found in glaciers is of widely varying sizes—from the finest rock particles to large boulders—and also generally of varied types as it includes the different rocks that a glacier is overriding. For this reason, a glacially abraded surface usually bears many different "tool-marks," from microscopic scratches to gouges centimetres deep and tens of metres long. Over thousands of years glaciers may erode their substrate to a depth of several tens of metres by this mechanism, producing a variety of streamlined landforms typical of glaciated landscapes.

Several other processes of glacial erosion are generally included under the terms glacial plucking or quarrying. This process involves the removal of larger pieces of rock from the glacier bed. Various explanations for this phenomenon have been proposed. Some of the mechanisms suggested are based on differential stresses in the rock caused by ice being forced to flow around bedrock obstacles. High stress gradients are particularly important, and the resultant tensile stresses can pull the rock apart along preexisting joints or crack systems. These pressures have been shown to be sufficient to fracture solid rock, thus making it available for removal by the ice flowing above it. Other possibilities include the forcing apart of rock by the pressure of crystallization produced beneath the glacier as water derived from the ice refreezes (regelation) or because of temperature fluctuations in cavities under the glacier. Still another possible mechanism involves hydraulic pressures of flowing water known to be present, at least temporarily, under nearly all temperate glaciers. It is hard to determine which process is dominant because access to the base of active glaciers is rarely possible. Nonetheless, investigators know that larger pieces of rock are plucked from the glacier bed and contribute to the number of abrasive "tools" available to the glacier at its base. Other sources for the rock debris in glacier ice may include rockfalls from steep slopes bordering a glacier or unconsolidated sediments overridden as a glacier advances.

GLACIAL DEPOSITION

Debris in the glacial environment may be deposited directly by the ice (till) or, after reworking, by meltwater streams (outwash). The resulting deposits are termed glacial drift.

As the ice in a valley glacier moves from the area of accumulation to that of ablation, it acts like a conveyor belt, transporting debris located beneath, within, and above the glacier toward its terminus or, in the case of an ice sheet, toward the outer margin. Near the glacier margin where the ice velocity decreases greatly is the zone of deposition. As the ice melts away, the debris that was originally frozen into the ice commonly forms a rocky and/or muddy blanket over the glacier margin. This layer often slides off the ice in the form of mudflows. The resulting deposit is called a flow-till by some authors. On the other hand, the debris may be laid down more or less in place as the ice melts away around and beneath it. Such deposits are referred to as melt-out till, and sometimes as ablation till. In many cases, the material located between a moving glacier and its bedrock bed is severely sheared, compressed, and "over-compacted." This type of deposit is called lodgment till.

By definition, till is any material laid down directly or reworked by a glacier. Typically, it is a mixture of rock fragments and boulders in a fine-grained sandy or muddy matrix (non-stratified drift). The exact composition of any particular till, however, depends on the materials available to the glacier at the time of deposition. Thus, some tills are made entirely of lake clays deformed by an overriding glacier. Other tills are composed of river gravels and sands that have been "bulldozed" and striated during a glacial advance. Tills often contain some of the tools that glaciers use to abrade their bed. These rocks and boulders bear striations, grooves, and facets, and characteristic till-stones are commonly shaped like bullets or flat-irons. Till-boulders of a rock type different from the bedrock on which they are deposited are dubbed "erratics." In some cases, erratics with distinctive lithologies can be traced back to their source, enabling investigators to ascertain the direction of ice movement of ice sheets in areas where striations either are absent or are covered by till or vegetation.

Meltwater deposits, also called glacial outwash, are formed in channels directly beneath the glacier or in lakes and streams in front of its margin. In contrast to till, outwash is generally bedded or laminated (stratified drift), and the individual layers are relatively well sorted according to grain size. In most cases, gravels and boulders in outwash are rounded and do not bear striations or grooves on their surfaces, since these tend to wear off rapidly during stream transport. The grain size of individual deposits depends not only on the availability of different sizes of debris but also on the velocity of the depositing current and the distance from the head of the stream. Larger boulders are deposited by rapidly flowing creeks and rivers close to the glacier margin. Grain size of deposited material decreases with increasing distance from the glacier. The finest fractions, such as clay and silt, may be deposited in glacial lakes or ponds or transported all the way to the ocean.

A swath of till in Glacier National Park, Montana. What type of till is left behind depends on the composition of each glacier, but typical materials are stone and silt. Department of Interior/USGS

Finally, it must be stressed that most glacier margins are constantly changing chaotic masses of ice, water, mud, and rocks. Ice-marginal deposits thus are of a highly variable nature over short distances, as is much the case with till and outwash as well.

EROSIONAL LANDFORMS

A number of landforms result from the movement of glaciers across Earth's surface. Such features in present-day glaciers range from rock polish and glacial grooves on smaller scales to hanging valleys and drumlins on larger ones. Most recently during the Pleistocene Epoch (about 2.6 million to 11,700 years ago), vast continental ice sheets reached into the middle latitudes. Most scientists maintain that these ice sheets carved out many of the world's present-day freshwater basins.

SMALL-SCALE FEATURES OF GLACIAL EROSION

Glacial erosion is caused by two different processes: abrasion and plucking. Nearly all glacially scoured erosional landforms bear the tool marks of glacial abrasion provided that they have not been removed by subsequent weathering. Even though these marks are not large enough to be called landforms, they constitute an integral part of any glacial landscape and thus warrant description here. The type of mark produced on a surface during glacial erosion depends on the size and shape of the tool, the pressure being applied to it, and the relative hardnesses of the tool and the substrate.

ROCK POLISH

The finest abrasive available to a glacier is the so-called rock flour produced by the constant grinding at the base of the ice. Rock flour acts like jewelers' rouge and produces microscopic scratches that, with time, smooth and polish rock surfaces, often to a high lustre.

STRIATIONS

Stirations are scratches visible to the naked eye, ranging in size from fractions of a millimetre to a few millimetres deep and a few millimetres to centimetres long. Large striations produced by a single tool may be several centimetres deep and wide and tens of metres long.

Because the striation-cutting tool was dragged across the rock surface by the ice, the long axis of a striation indicates the direction of ice movement in the immediate vicinity of that striation. Determination of the regional direction of movement of former ice sheets, however, requires measuring hundreds of striation directions over an extended area because ice moving close to the base of a glacier is often locally deflected by bedrock obstacles. Even when such a regional study is conducted, additional information is frequently needed in low-relief areas to determine which end of the striations points down-ice toward the former outer margin of the glacier. On an outcrop scale, such information can be gathered by studying "chatter marks." These crescentic gouges and lunate fractures are caused by the glacier dragging a rock or boulder over a hard and brittle rock surface and forming a series of sickle-shaped gouges. Such depressions in the bedrock are steep-sided on their "up-glacier" face and have a lower slope on their down-ice side. Depending on whether the horns of the sickles point up the glacier or down it, the chatter marks are designated crescentic gouges or lunate fractures.

Another small-scale feature that allows absolute determination of the direction in which the ice moved is what is termed knob-and-tail. A knob-and-tail is formed during glacial abrasion of rocks that locally contain spots more resistant than the surrounding rock, as is the case, for example, with silicified fossils in limestone. After abrasion has been active for some time, the harder parts of the rock form protruding knobs as the softer rock is preferentially eroded away around them. During further erosion, these protrusions protect the softer rock on their lee side and a tail forms there, pointing from the knob to the margin of the glacier. The scale of these features depends primarily on the size of the inhomogeneities in the rock and ranges from fractions of millimetres to metres.

P-FORMS AND GLACIAL GROOVES

These features, which extend several to tens of metres in length, are of uncertain origin. P-forms (P for plastically molded) are smooth-walled, linear depressions which may be straight, curved, or sometimes hairpin-shaped and measure tens of centimetres to metres in width and depth. Their cross sections are often semicircular to parabolic, and their walls are commonly striated parallel to their long axis, indicating that ice once flowed in them. Straight P-forms are frequently called glacial grooves, even though the term is also applied to large striations, which, unlike the P-forms, were cut by a single tool. Some researchers believe that P-forms were not carved directly by the ice but rather were eroded by pressurized mud slurries flowing beneath the glacier.

THE EROSIONAL LANDFORMS OF VALLEY GLACIERS

Many of the world's higher mountain ranges—such as the Alps, the North and South American Cordilleras, the Himalayas, and the Southern Alps in New Zealand, as well as the mountains of Norway, including those of

Spitsbergen—are partly glaciated today. During periods of the Pleistocene, such glaciers were greatly enlarged and filled most of the valleys with ice, even reaching far beyond the mountain front in certain places. Most scenic alpine landscapes featuring sharp mountain peaks, steep-sided valleys, and innumerable lakes and waterfalls are a product of several periods of glaciation.

Erosion is generally greater than deposition in the upper reaches of a valley glacier, whereas deposition exceeds erosion closer to the terminus. Accordingly, erosional landforms dominate the landscape in the high areas of glaciated mountain ranges.

CIRQUES, TARNS, U-SHAPED VALLEYS, ARÊTES, AND HORNS

The heads of most glacial valleys are occupied by one or several cirques (or corries). A cirque is an amphitheatre-shaped hollow with the open end facing down-valley. The back is formed by an arcuate cliff called the headwall. In an ideal cirque, the headwall is semicircular in plan view. This situation, however, is generally found only in cirques cut into flat plateaus. More common are headwalls angular in map view due to irregularities in height along their perimeter. The bottom of many cirques is a shallow basin, which may contain a lake. This basin and the base of the adjoining headwall usually show signs of extensive glacial abrasion and plucking.

Even though the exact process of cirque formation is not entirely understood, it seems that the part of the headwall above the glacier retreats by frost shattering and ice wedging. The rock debris then falls either onto the surface of the glacier or into the randkluft or bergschrund. Both names describe the crevasse between the ice at the head of the glacier and the cirque headwall. The rocks on

A glacial lake, or tarn, found in a cirque of the Pioneer Mountains, Montana. The glacial headwall rises up toward the back of this natural amphitheatre. Dr. Marli Miller/Visuals Unlimited/Getty Images

the surface of the glacier are successively buried by snow and incorporated into the ice of the glacier. Because of a downward velocity component in the ice in the accumulation zone, the rocks are eventually moved to the base of the glacier. At that point, these rocks, in addition to the rock debris from the bergschrund, become the tools with which the glacier erodes, striates, and polishes the base of the headwall and the bottom of the cirque.

During the initial growth and final retreat of a valley glacier, the ice often does not extend beyond the cirque. Such a cirque glacier is probably the main cause for the formation of the basin scoured into the bedrock bottom of many cirques. Sometimes these basins are "over-deepened" several tens of metres and contain lakes called tarns.

In contrast to the situation in a stream valley, all debris falling or sliding off the sides and the headwalls of a glaciated valley is immediately removed by the flowing ice. Moreover, glaciers are generally in contact with a much larger percentage of a valley's cross section than equivalent rivers or creeks. Thus glaciers tend to erode the bases of the valley walls to a much greater extent than do streams, whereas a stream erodes an extremely narrow line along the lowest part of a valley. The slope of the adjacent valley walls depends on the stability of the bedrock and the angle of repose of the weathered rock debris accumulating at the base of and on the valley walls. For this reason, rivers tend to form V-shaped valleys. Glaciers, which inherit V-shaped stream valleys, reshape them drastically by first removing all loose debris along the base of the valley walls and then preferentially eroding the bedrock along the base and lower sidewalls of the valley. In this way, glaciated valleys assume a characteristic parabolic or U-shaped cross profile, with relatively wide and flat bottoms and steep, even vertical sidewalls.

By the same process, glaciers tend to narrow the bedrock divides between the upper reaches of neighbouring parallel valleys to jagged, knife-edge ridges known as *arêtes*, which also form between two cirques facing in opposite directions. The low spot, or saddle, in the arête between two cirques is called a col. A higher mountain often has three or more cirques arranged in a radial pattern on its flanks. Headward erosion of these cirques finally leaves only a sharp peak flanked by nearly vertical headwall cliffs, which are separated by arêtes. Such glacially eroded mountains are termed horns, the most widely known of which is the Matterhorn in the Swiss Alps.

HANGING VALLEYS

Large valley glacier systems consist of numerous cirques and smaller valley glaciers that feed ice into a large trunk glacier. Because of its greater ice discharge, the trunk glacier has greater erosive capability in its middle and lower reaches than smaller tributary glaciers that join it there. The main valley is therefore eroded more rapidly than the side valleys. With time, the bottom of the main valley becomes lower than the elevation of the tributary valleys. When the ice has retreated, the tributary valleys are left joining the main valley at elevations substantially higher than its bottom. Tributary valleys with such unequal or discordant junctions are called hanging valleys. In extreme cases where a tributary joins the main valley high up in the steep part of the U-shaped trough wall, waterfalls may form after deglaciation, as in Yosemite and Yellowstone national parks in the western United States.

PATERNOSTER LAKES

Some glacial valleys have an irregular, longitudinal bedrock profile, with alternating short, steep steps and longer, relatively flat portions. Even though attempts have been made to explain this feature in terms of some inherent characteristic of glacial flow, it seems more likely that differential erodibility of the underlying bedrock is the real cause of the phenomenon. Thus the steps are probably formed by harder or less fractured bedrock, whereas the flatter portions between the steps are underlain by softer or more fractured rocks. In some cases, these softer areas have been excavated by a glacier to form shallow bedrock basins. If several of these basins are occupied by lakes along one glacial trough in a pattern similar to beads on a string, they are called *paternoster* (Latin: "our father") lakes by analogy with a string of rosary beads.

ROCHES MOUTONNÉES

These structures are bedrock knobs or hills that have a gently inclined, glacially abraded, and streamlined stoss side (i.e., one that faces the direction from which the over-riding glacier impinged) and a steep, glacially plucked lee side. They are generally found where jointing or fracturing in the bedrock allows the glacier to pluck the lee side of the obstacle. In plan view, their long axes are often, but not always, aligned with the general direction of ice movement.

ROCK DRUMLINS

A feature similar to roches moutonnées, rock drumlins are bedrock knobs or hills completely streamlined, usually with steep stoss sides and gently sloping lee sides. Both roches moutonnées and rock drumlins range in length from several metres to several kilometres and in height from tens of centimetres to hundreds of metres. They are typical of both valley and continental glaciers. The larger ones, how-ever, are restricted to areas of continental glaciation.

THE EROSIONAL LANDFORMS OF CONTINENTAL GLACIERS

In contrast to valley glaciers, which form exclusively in areas of high altitude and relief, continental glaciers, including the great ice sheets of the past, occur in high and middle latitudes in both hemispheres, covering landscapes that range from high alpine mountains to low-lying areas with negligible relief. Therefore, the landforms produced by continental glaciers are more diverse and widespread. Yet, just like valley glaciers, they have an area where ero-sion is the dominant process and an area close to their margins where net deposition generally occurs.

The capacity of a continental glacier to erode its substrate has been a subject of intense debate. All of the areas formerly covered by ice sheets show evidence of areally extensive glacial scouring. The average depth of glacial erosion during the Pleistocene probably did not exceed a few tens of metres, however. This is much less than the deepening of glacial valleys during mountain glaciation. One of the reasons for the apparent limited erosional capacity of continental ice sheets in areas of low relief may be the scarcity of tools available to them in these regions. Rocks cannot fall onto a continental ice sheet in the accumulation zone, because the entire landscape is buried. Thus, all tools must be quarried by the glacier from the underlying bedrock. With time, this task becomes increasingly difficult as bedrock obstacles are abraded and streamlined. Nonetheless, the figure for depth of glacial erosion during the Pleistocene cited above is an average value, and locally several hundreds of metres of bedrock were apparently removed by the great ice sheets. Such enhanced erosion seems concentrated at points where the glaciers flowed from hard, resistant bedrock onto softer rocks or where glacial flow was channelized into outlet glaciers.

As a continental glacier expands, it strips the underlying landscape of the soil and debris accumulated at the preglacial surface as a result of weathering. The freshly exposed harder bedrock is then eroded by abrasion and plucking. During this process, bedrock obstacles are shaped into streamlined "whaleback" forms, such as roches moutonnées and rock drumlins. The adjoining valleys are scoured into rock-floored basins with the tools plucked from the lee sides of roches moutonnées. The long axes of the hills and valleys are often preferentially oriented in the direction of ice flow. An area totally composed of smooth whaleback forms and basins is called a streamlined landscape.

Streams cannot erode deep basins because water cannot flow uphill. Glaciers, on the other hand, can flow uphill over obstacles at their base as long as there is a sufficient slope on the upper ice surface pointing in that particular direction. Therefore the great majority of the innumerable lake basins and small depressions in formerly glaciated areas can only be a result of glacial erosion. Many of these lakes, such as the Finger Lakes in the U.S. state of New York, are aligned parallel to the direction of regional ice flow. Other basins seem to be controlled by preglacial drainage systems. Yet, other depressions follow the structure of the bedrock, having been preferentially scoured out of areas underlain by softer or more fractured rock.

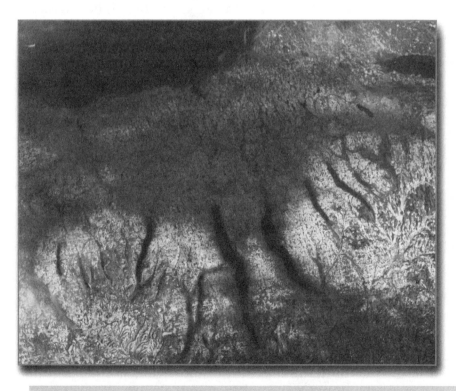

An aerial view of New York's Finger Lakes, taken from the International Space Station. Speculation is that the lake basins were formed by ice sheets as much as 3 km (1.9 miles) deep. NASA

A number of the largest freshwater lake basins in the world, such as the U.S. Great Lakes or the Great Slave Lake and Great Bear Lake in Canada, are situated along the margins of the Precambrian shield of North America. Many researchers believe that glacial erosion was especially effective at these locations because the glaciers could easily abrade the relatively soft sedimentary rocks to the south with hard, resistant crystalline rocks brought from the shield areas that lie to the north. Nonetheless, further research is necessary to determine how much of the deepening of these features can be ascribed to glacial erosion, as opposed to other processes such as tectonic activity or preglacial stream erosion.

Fjords are found along some steep, high-relief coastlines where continental glaciers formerly flowed into the sea. They are deep, narrow valleys with U-shaped cross sections that often extend inland for tens or hundreds of kilometres and are now partially drowned by the ocean. These troughs are typical of the Norwegian coast, but they also are found in Canada, Alaska, Iceland, Greenland, Antarctica, New Zealand, and southernmost Chile. The floor and steep walls of fjords show ample evidence of glacial erosion. The long profile of many fjords, including alternating basins and steps, is very similar to that of glaciated valleys. Toward the mouth, fjords may reach great depths, as in the case of Sogn Fjord in southern Norway where the maximum water depth exceeds 1,300 metres (about 4,300 feet). At the mouth of a fjord, however, the floor rises steeply to create a rock threshold, and water depths decrease markedly. At Sogn Fjord the water at this "threshold" is only 150 metres (about 500 feet) deep, and in many fjords the rock platform is covered by only a few metres of water.

The exact origin of fjords is still a matter of debate. While some scientists favour a glacial origin, others believe that much of the relief of fjords is a result of tectonic activity and that glaciers only slightly modified preexisting large valleys. In order to erode Sogn Fjord to its present depth, the glacier occupying it during the maximum of the Pleistocene must have been 1,800 to 1,900 metres (5,900 to 6,200 feet) thick. Such an ice thickness may seem extreme, but even now, during an interglacial period, the Skelton Glacier in Antarctica has a maximum thickness of about 1,450 metres (about 4,750 feet). This outlet glacier of the Antarctic ice sheet occupies a trough, which in places is more than 1 km (0.6 mile) below sea level and would become a fjord in the event of a large glacial retreat.

DEPOSITIONAL LANDFORMS

The movement of glaciers across terrestrial environments disturbs the underlying layers of soil and rock, eventually delivering and depositing the collected material to the end, or margin, of the glacier. As a glacier moves along a valley, it picks up rock debris from the valley walls and floor, transporting it in, on, or under the ice. This scouring process also occurs in continental glaciers. When the material reaches the lower parts of the glacier where ablation is dominant, it is concentrated along the glacier margins as more and more debris melts out of the ice. If the position of the glacier margin is constant for an extended amount of time, larger accumulations of glacial debris (till) will form at the glacier margin. In addition, a great deal of material is rapidly flushed through and out of the glacier by meltwater streams flowing under, within, on, and next to the glacier. Part of this streamload is deposited in front of the glacier close to its snout, or terminus. There, it may mix with material brought by, and melting out from, the glacier as well as with material washed in from other, nonglaciated tributary valleys.

THE DEPOSITIONAL LANDFORMS OF
VALLEY GLACIERS

In contrast to continental glaciers, valley glaciers are con-
stricted within areas of high elevation, and thus rock, soil,
and debris taken up typically originates in the valley in
which the glacier has formed. Material pushed ahead of
the glacier or engulfed by its forward edge often forms a
moraine, whereas pressurized mud that develops underneath the glacier may be used to construct flutes under
the right conditions.

MORAINES

After the initial wave of material is deposited at the terminus of a glacier, the glacier itself may continue to
advance or readvance after a period of retreat. This activity effectively "bulldozes" all the loose material in front of
it into a ridge of chaotic debris that closely hugs the shape
of the glacier snout. Any such accumulation of till melted
out directly from the glacier or piled into a ridge by the
glacier is a moraine. Large valley glaciers are capable of
forming moraines a few hundred metres high and many
hundreds of metres wide. Linear accumulations of till
formed immediately in front of or on the lower end of the
glacier are end moraines. The moraines formed along
the valley slopes next to the side margins of the glacier are
termed lateral moraines.

During a single glaciation, a glacier may form many
such moraine arcs, but all the smaller moraines, which
may have been produced during standstills or short
advances while the glacier moved forward to its outer-
most ice position, are generally destroyed as the glacier
resumes its advance. The end moraine of largest extent
formed by the glacier (which may not be as extensive as
the largest ice advance) during a given glaciation is called

the terminal moraine of that glaciation. Successively smaller moraines formed during standstills or small readvances as the glacier retreats from the terminal moraine position are recessional moraines, and moraines that form when two glaciers meet one another are called medial moraines.

FLUTES

The depositional equivalent of erosional knob-and-tail structures are known as flutes. Close to the lower margin, some glaciers accumulate so much debris beneath them that they actually glide on a bed of pressurized muddy till. As basal ice flows around a pronounced bedrock knob or a boulder lodged in the substrate, a cavity often forms in the ice on the lee side of the obstacle because of the high viscosity of the ice. Any pressurized muddy paste present under the glacier may then be injected into this cavity and deposited as an elongate tail of till, or flute. The size depends mainly on the size of the obstacle and on the availability of subglacial debris. Flutes vary in height from a few centimetres to tens of metres and in length from tens of centimetres to kilometres, even though very large flutes are generally limited to continental ice sheets.

THE DEPOSITIONAL LANDFORMS OF CONTINENTAL GLACIERS

Many of the deposits of continental ice sheets are very similar to those of valley glaciers. Terminal, end, and recessional moraines are formed by the same process as with valley glaciers, but they can be much larger. Morainic ridges may be laterally continuous for hundreds of kilometres, hundreds of metres high, and several kilometres wide. Since each moraine forms at a discreet position of the ice margin, plots of end moraines on a map of suitable scale allow the reconstruction of ice sheets at varying stages during their retreat.

In addition to linear accumulations of glacial debris, continental glaciers often deposit a more or less continuous, thin (less than 10 metres [33 feet]) sheet of till over large areas, which is called ground moraine. This type of moraine generally has a "hummocky" topography of low relief, with alternating small till mounds and depressions. Swamps or lakes typically occupy the low-lying areas. Flutes are a common feature found in areas covered by ground moraine.

Another depositional landform associated with continental glaciation is the drumlin, a streamlined, elongate mound of sediment. Such structures often occur in groups of tens or hundreds, which are called drumlin fields. The long axis of individual drumlins is usually aligned parallel to the direction of regional ice flow. In long profile, the stoss side of a drumlin is steeper than the lee side. Some drumlins consist entirely of till, while others have bedrock cores draped with till. The till in many drumlins has been shown to have a "fabric" in which the long axes of the individual rocks and sand grains are aligned parallel to the ice flow over the drumlin. Even though the details of the process are not fully understood, drumlins seem to form subglacially close to the edge of an ice sheet, often directly down-ice from large lake basins overridden by the ice during an advance. The difference between a rock drumlin and a drumlin is that the former is an erosional bedrock knob, whereas the latter is a depositional till feature.

MELTWATER DEPOSITS

Much of the debris in the glacial environment of both valley and continental glaciers is transported, reworked, and laid down by water. Whereas glaciofluvial deposits are formed by meltwater streams, glaciolacustrine sediments accumulate at the margins and bottoms of glacial lakes and ponds.

GLACIOFLUVIAL DEPOSITS

The discharge of glacial streams is highly variable, depending on the season, time of day, and cloud cover. Maximum discharges occur during the afternoon on warm, sunny summer days, and minima on cold winter mornings. Beneath or within a glacier, the water flows in tunnels and is generally pressurized during periods of high discharge. In addition to debris washed in from unglaciated highlands adjacent to the glacier, a glacial stream can pick up large amounts of debris along its path at the base of the glacier. For this reason, meltwater streams issuing forth at the snout of a valley glacier or along the margin of an ice sheet are generally laden to transporting capacity with debris.

Beyond the glacier margin, the water, which is no longer confined by the walls of the ice tunnel, spreads out and loses some of its velocity. Because of the decreased velocity, the stream must deposit some of its load. As a result, the original stream channel is choked with sediments, and the stream is forced to change its course around the obstacles, often breaking up into many winding and shifting channels separated by sand and gravel bars. The highly variable nature of the sediments laid down by such a braided stream reflects the unstable environment in which they form. Lenses of fine-grained, cross-bedded sands are often interbedded laterally and vertically with stringers of coarse, bouldery gravel. Since the amount of sediment laid down generally decreases with distance from the ice margin, the deposit is often wedge-shaped in cross section, ideally gently sloping off the end moraine formed at that ice position and thinning downstream. The outwash is then said to be "graded to" that particular moraine.

In map view, the shape of the deposit depends on the surrounding topography. Where the valleys are deep enough not to be buried by the glaciofluvial sediments, as in most mountainous regions, the resulting elongate, planar deposits are termed valley trains. On the other hand, in low-relief areas the deposits of several ice-marginal streams may merge to form a wide outwash plain, or sandur.

If the ice margin stabilizes at a recessional position during glacial retreat, another valley train or sandur may be formed inside of the original one. Because of the downstream thinning of the outwash at any one point in the valley, the recessional deposit will be lower than and inset into the outer, slightly older outwash plain. Flat-topped remnants of the older plain may be left along the valley sides; these are called terraces. Ideally each recessional ice margin has a terrace graded to it, and these structures can be used in addition to moraines to reconstruct the positions of ice margins through time. In some cases where the glacier either never formed moraines or where the moraines were obliterated by the outwash or postglacial erosion, terraces are the only means of ice margin reconstruction.

Streams that flow over the terminus of a glacier often deposit stratified drift in their channels and in depressions on the ice surface. As the ice melts away, this ice-contact stratified drift slumps and partially collapses to form stagnant ice deposits. Isolated mounds of bedded sands and gravels deposited in this manner are called kames. Kame terraces form in a similar manner but between the lateral margin of a glacier and the valley wall. Glacial geologists sometimes employ the term kame moraine to describe deposits of stratified drift laid down at an ice margin in the arcuate shape of a moraine. Some researchers, however, object to the use of the term moraine in this context because the deposit is not composed of till.

In some cases, streams deposit stratified drift in sub-glacial or englacial tunnels. As the ice melts away, these sinuous channel deposits may be left as long linear gravel ridges called eskers. Some eskers deposited by the great ice sheets of the Pleistocene can be traced for hundreds of kilometres, even though most esker segments are only a few hundred metres to kilometres long and a few to tens of metres high.

Kettles, potholes, or ice pits are steep-sided depressions typical of many glacial and glaciofluvial deposits. Kettles form when till or outwash is deposited around ice blocks that have become separated from the active glacier by ablation. Such "stagnant" ice blocks may persist insulated under a mantle of debris for hundreds of years. When they finally melt, depressions remain in their place, bordered by slumped masses of the surrounding glacial deposits. Many of the lakes in areas of glacial deposition are water-filled kettles and so are called kettle lakes. If a sandur or valley train contains many kettles, it is referred to as a pitted outwash plain.

GLACIOLACUSTRINE DEPOSITS

Glacial and proglacial lakes are found in a variety of environments and in considerable numbers. Erosional lake basins have already been mentioned, but many lakes are formed as streams are dammed by the ice itself, by glacial deposits, or by a combination of these factors. Any lake that remains at a stable level for an extended period of time (such as hundreds or thousands of years) tends to form a perfectly horizontal, flat, terracelike feature along its beach. Such a bench may be formed by wave erosion of the bedrock or glacial sediments that form the margin of the lake, and it is called a wave-cut bench. On the other hand, it may be formed by deposition of sand and gravel from long-shore currents along the margin of the lake, in which case it is referred to as a beach ridge. The width of these shorelines varies from a few metres to several hundred metres. As the lake level is lowered due to the opening of another outlet or downcutting of the spillway, new, lower shorelines may be formed. Most former or existing glacial lakes (such as the Great Salt Lake and the Great Lakes in North America) have several such shorelines that can be used both to determine the former size and depth of now-extinct or shrunken lakes and to determine the amount of differential postglacial uplift because they are now tilted slightly from their original horizontal position.

Where a stream enters a standing body of water, it is forced to deposit its bedload. The coarser gravel and sand are laid down directly at the mouth of the stream as successive, steeply inclined foreset beds. The finer, suspended silt and clay can drift a bit farther into the lake, where they are deposited as almost flat-lying bottomset beds. As the sediment builds out farther into the lake (or ocean), the river deposits a thin veneer of subhorizontal gravelly topset beds over the foreset units. Because the foreset–topset complex often has the shape of a triangle with the mouth of the stream at one apex, such a body of sediment is called a delta. Many gravel and sand pits are located in deltas of former glacial lakes.

The flat-lying, fine-grained bottomset beds of many large former glacial lakes filled in and buried all of the pre-existing relief and are now exposed, forming perfectly flat lake plains. Cuts into these sediments often reveal rhythmically interbedded silts and clays. Some of these so-called rhythmites have been shown to be the result of seasonal changes in the proglacial environment. During the warmer summer months, the meltwater streams carry silt and clay into the lakes, and the silt settles out of suspension more rapidly than the clay. A thicker, silty summer layer is thus deposited. During the winter, as the surface of the lake freezes and the meltwater discharge into it ceases, the clays contained in the lake water slowly settle out of suspension to form a thin winter clay layer. Such lacustrine deposits with annual silt and clay "couplets" are known as varves.

PERIGLACIAL LANDFORMS

In the cold, or periglacial (near-glacial), areas adjacent to and beyond the limit of glaciers, a zone of intense freeze-thaw activity produces periglacial features and landforms. This happens because of the unique behaviour of water as it changes from the liquid to the solid state. As water freezes, its volume increases about 9 percent. This is often combined with the process of differential ice growth, which traps air, resulting in an even greater increase in volume. If confined in a crack or pore space, such ice and air mixtures can exert pressures of about 200,000 kilopascals (29,000 pounds per square inch). This is enough to break the enclosing rock. Thus freezing water can be a powerful agent of physical weathering. If multiple freeze-and-thaw cycles occur, the growth of ice crystals fractures and moves material by means of frost shattering and frost heaving, respectively. In addition, in permafrost regions where the ground remains frozen all year, characteristic landforms are formed by perennial ice.

FELSENMEERS, TALUS, AND ROCK GLACIERS

In nature, the tensional strength of most rocks is exceeded by the pressure of water crystallizing in cracks. Thus, repeated freezing and thawing not only forms potholes in poorly constructed roads but also is capable of reducing exposed bedrock outcrops to rubble. Many high peaks are covered with frost-shattered angular rock fragments. A larger area blanketed with such debris is called a *felsenmeer*, from the German for "sea of rocks." The rock fragments can be transported downslope by flowing water or frost-induced surface creep, or they may fall off the cliff from which they were wedged by the ice. Accumulations of this angular debris at the base of steep slopes are known as talus. Owing to the steepness of the valley sides of many glacial troughs, talus is commonly found in formerly glaciated mountain regions. Talus cones are formed when the debris coming from above is channelized on its way to the base of the cliff in rock chutes. As the talus cones of neighbouring chutes grow over time, they may coalesce to form a composite talus apron.

In higher mountain regions, the interior of thick accumulations of talus may remain at temperatures below freezing all year. Rain or meltwater percolating into the interstices between the rocks freezes over time, filling the entire pore space. In some cases, enough ice forms to enable the entire mass of rock and ice to move downhill like a glacier. The resulting massive, lobate, mobile feature is called a rock glacier. Some rock glaciers have been shown to contain pure ice under a thick layer of talus with some interstitial ice. These features may be the final retreat stages of valley glaciers buried under talus.

CHAPTER 6

ICEBERGS AND SEA ICE

Although icebergs and sea ice often appear together, they are very different phenomena. Icebergs are floating masses of freshwater ice that have broken off from the seaward end of either a glacier or an ice shelf. They are found in the oceans surrounding Antarctica, in the seas of the Arctic and subarctic, in Arctic fjords, and in lakes fed by glaciers. In contrast, sea ice is essentially frozen seawater. It also occurs within the seas surrounding Antarctica and the Arctic Ocean, as well as its adjacent seas as far south as China and Japan.

ICEBERGS

Icebergs are magnificent structures composed of a freeboard section (height above the waterline) and an underwater section. Their size is greatest when they initially calve, and Arctic bergs tend to be much smaller than the largest Antarctic ones. After they calve from their parent glacier or ice shelf, icebergs may be released into the open ocean, where they slowly erode from wave action or melt from rising environmental temperatures.

THE ORIGIN OF ANTARCTIC ICEBERGS

Icebergs of the Antarctic calve from floating ice shelves and are a magnificent sight, forming huge, flat "tabular" structures. A typical newly calved iceberg of this type has a diameter that ranges from several kilometres to tens of kilometres, a thickness of 200–400 metres (660–1,320 feet), and a freeboard, or the height of the "berg" above the waterline, of 30–50 metres (100–160 feet). The mass of a tabular iceberg is typically several billion tons. Floating ice shelves are a continuation of the flowing mass of ice that makes up the continental ice sheet. Floating ice shelves fringe about 30 percent of Antarctica's coastline. The transition area where floating ice meets ice that sits directly on bedrock is known as the grounding line.

Under the pressure of the ice flowing outward from the centre of the continent, the ice in these shelves moves seaward at 0.3–2.6 km (0.2–1.6 miles) per year. The exposed seaward front of the ice shelf experiences stresses from subshelf currents, tides, and ocean swell in the summer and moving pack ice during the winter. Since the shelf normally possesses cracks and crevasses, it will eventually fracture to yield freely floating icebergs. Some minor ice shelves generate large iceberg volumes because of their rapid velocity; the small Amery Ice Shelf, for instance, produces 31 cubic km (about 7 cubic miles) of icebergs per year as it drains about 12 percent of the east Antarctic Ice Sheet.

Iceberg calving may be caused by ocean wave action, contact with other icebergs, or the behaviour of melting water on the upper surface of the berg. With the use of tiltmeters (tools that can detect a change in the angle of the slope of an object), scientists monitoring iceberg-calving events have been able to link the breaking stress occurring near the ice front to long storm-generated swells originating tens of thousands of kilometres away. This bending stress is enhanced in the case of glacier tongues (long narrow floating ice shelves produced by fast-flowing glaciers that protrude far into the ocean). The swell causes the tongue to oscillate until it fractures. In addition, on a

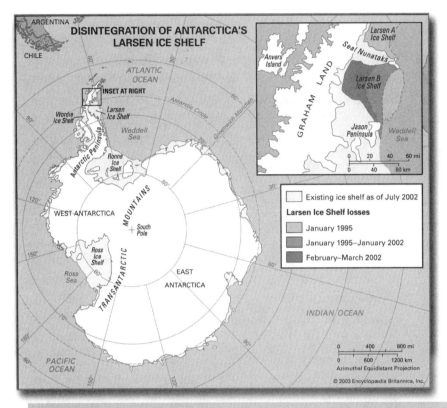

Map showing the extent of collapse of the Larsen Ice Shelf. The Larsen A Ice Shelf disintegrated in 1995, whereas the Larsen B Ice Shelf broke apart in 2002. Both events were caused by water from surface melting that ran down into crevasses, refroze, and wedged each shelf into pieces.

number of occasions, iceberg calving has been observed immediately after the collision of another iceberg with the ice front. Furthermore, the mass breakout of icebergs from the Larsen Ice Shelf between 1995 and 2002, though generally ascribed to global warming, is thought to have occurred because summer meltwater on the surface of the shelf filled nearby crevasses. As the liquid water refroze, it expanded and produced fractures at the bases of the crevasses. This phenomenon, known as frost wedging, caused the shelf to splinter in several places and brought about the disintegration of the shelf.

THE ORIGIN OF ARCTIC ICEBERGS

Most Arctic icebergs originate from the fast-flowing glaciers that descend from the Greenland Ice Sheet. Many
glaciers are funneled through gaps in the chain of coastal
mountains. The irregularity of the bedrock and valley wall
topography both slows and accelerates the progress of glaciers. These stresses cause crevasses to form, which are
then incorporated into the structure of the icebergs.
Arctic bergs tend to be smaller and more randomly shaped
than Antarctic bergs and also contain inherent planes of
weakness, which can easily lead to further fracturing. If
their draft exceeds the water depth of the submerged sill
at the mouth of the fjord, newly calved bergs may stay
trapped for long periods in their fjords of origin. Such an
iceberg will change shape, especially in summer as the
water in the fjord warms, through the action of differential melt rates occurring at different depths. Such
variations in melting can affect iceberg stability and cause
the berg to capsize. Examining the profiles of capsized
bergs can help researchers detect the variation of summer
temperature occurring at different depths within the
fjord. In addition, the upper surfaces of capsized bergs
may be covered by small scalloped indentations that are
by-products of small convection cells that form when ice
melts at the ice-water interface.

The Arctic Ocean's equivalent of the classic tabular
iceberg of Antarctic waters is the ice island. Ice islands can
be up to 30 km (19 miles) long but are only some 60 metres
(200 feet) thick. The main source of ice islands used to be
the Ward Hunt Ice Shelf on Canada's Ellesmere Island
near northwestern Greenland, but the ice shelf has been
retreating as ice islands and bergs continue to calve from
it. (The ice shelf is breaking into pieces faster than new ice

can be formed.) Since the beginning of observations in the 1950s, the Ward Hunt Ice Shelf has virtually disappeared. The most famous of its ice islands was T-3, which was so named because it was the third in a series of three radar targets detected north of Alaska. This ice island carried a manned scientific station from 1952 to 1974. Ice islands produced by Ellesmere Island calve into the Beaufort Gyre (the clockwise-rotating current system in the Arctic Ocean) and may make several circuits of the Canada Basin before exiting the Arctic Ocean via Fram Strait (an ocean passage between Svalbard and Greenland).

Mountain peaks project through the ice cap on northern Ellesmere Island, Canada. Fred Bruemmer

A third source of ice islands, one that has become more active, is northeastern Greenland. The Flade Isblink, a small ice cap on Nordostrundingen in the northeastern corner of Greenland, calves thin tabular ice islands with clearly defined layering into Fram Strait. Observations in 1984 showed 60 grounded bergs with freeboards of 12–15 metres (40–50 feet) off Nordostrundingen in 37–53 metres (120–175 feet) of water. Similar bergs acted as pinning points for pressure ridges, which produced a blockage of the western part of Fram Strait for several years during the 1970s. In 2003 the multiyear cover of fast ice along the northeastern Greenland coast broke out. This allowed a huge number of tabular icebergs to emerge from the fast-flowing Nioghalvfjerdsfjorden Glacier and Zachariae Isstrøm in northeastern Greenland. Some of these reached the Labrador Sea two to three years later, while others remained grounded in 80–110 metres (260–360 feet) of water on the Greenland shelf.

ICEBERG STRUCTURE

A newly calved Antarctic tabular iceberg retains the physical properties of the outer part of the parent ice shelf. The shelf has the same layered structure as the continental ice sheet from which it flowed. All three features are topped with recently fallen snow that is underlain by older annual layers of increasing density. Annual layers are often clearly visible on the vertical side of a new tabular berg, which implies that the freeboard of the iceberg is mainly composed of compressed snow rather than ice. Density profiles through newly calved bergs show that at the surface of the berg the density might be only 400 kg per cubic metre (25 pounds per cubic foot)—pure ice has a density of 920 kg per cubic metre (57 pounds per cubic foot)—and both air and water may pass through the spaces between the crystal grains. Only when the density reaches 800 kg per cubic metre (50 pounds per cubic foot) deep within the berg do the air channels collapse to form air bubbles. At this point, the material can be properly classified as "ice," whereas the lower- density material above the ice is more properly called "firn."

Corresponding to a layer some 150–200 years old and coinciding approximately with the waterline, the firn-ice transition occurs about 40–60 metres (130–200 feet) below the surface of the iceberg. Deeper still, as density and pressure increase, the air bubbles become compressed. Within the Greenland Ice Sheet, pressures of 10–15 atmospheres (10,100–15,200 millibars) have been measured; the resulting air bubbles tend to be elongated, possessing lengths up to 4 mm (0.2 inch) and diameters of 0.02–0.18 mm (0.0008–0.007 inch). In Antarctic ice shelves and icebergs, the air bubbles are more often spherical or ellipsoidal and possess a diameter of 0.33–0.49 mm (0.01–0.02 inch). The size of the air bubbles decreases with increasing depth within the ice.

As soon as an iceberg calves, it starts to warm relative to its parent ice shelf. This warming accelerates as the berg drifts into more temperate regions, especially when it drifts free of the surrounding pack ice. Once the upper surface of the berg begins to melt, the section above the waterline warms relatively quickly to temperatures that approach the melting point of ice. Meltwater at the surface can percolate through the permeable uppermost 40–60 metres (130–200 feet) and refreeze at depth. This freezing releases the berg's latent heat, and the visible part of the berg becomes a warm mass that has little mechanical strength; it is composed of firn and thus can be easily eroded. The remaining mechanical strength of the iceberg

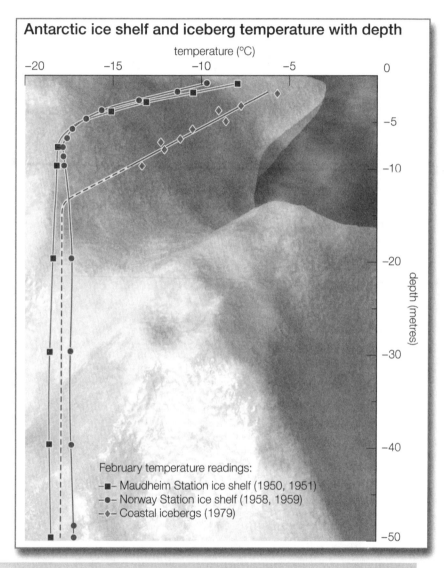

Antarctic ice shelf and iceberg temperature with depth

As ice depth increases to 12 metres (40 feet) and beyond, the temperature difference between icebergs and ice shelves is negligible. Encyclopædia Britannica, Inc.

is contained in the "cold core" below sea level, where temperatures remain at -15 to -20°C (5 to -4°F). In the cold core, heat transfer is inhibited owing to the lack of percolation and refreezing.

ICEBERG SIZE AND SHAPE

For many years, the largest reliably measured Antarctic iceberg was the one first observed off Clarence Island (one of the South Shetland Islands) by the whale catcher *Odd I* in 1927. It was 180 km (110 miles) long, was approximately square, and possessed a freeboard of 30–40 metres (100–130 feet). In 1956 an iceberg was sighted by USS *Glacier* off Scott Island (a small island about 500 km [300 miles] northeast of Victoria Land in the Ross Sea) with unconfirmed length of 335 km (210 miles) and width of 100 km (60 miles).

However, there have been many calvings of giant icebergs in the Ross and Weddell seas with dimensions that have been measured accurately by satellite. In 2000 iceberg B-15 broke off the Ross Ice Shelf with an initial length of 295 km (about 185 miles). Although B-15 broke into two fragments after a few days, B-15A—the larger portion, measuring 120 km (75 miles) long by 20 km (12 miles) wide—obstructed the entrance to McMurdo Sound and prevented the pack ice in the sound from clearing out in the summer. In October 2005 B-15A broke up into several large pieces off Cape Adare in Victoria Land because of the impact of distant swell. Iceberg C-19 was an even larger but narrow iceberg that broke off the Ross Ice Shelf in May 2002. It fragmented before it could drift far.

The Antarctic Peninsula has been warming significantly in recent decades (by 2.5 °C [4.5 °F] since the 1950s). Three ice shelves on the peninsula, the Wordie and Wilkins ice shelves on the west side of the peninsula and the Larsen Ice Shelf on the east side, have been disintegrating. This has caused the release of tremendous numbers of icebergs. The Larsen Ice Shelf has retreated twice since 2000; each event involved the fracture and release of a vast area of shelf ice in the form of multiple gigantic icebergs and innumerable smaller ones. The breakout of 3,250 square km (1,250 square miles) of shelf over 35 days in early 2002 effectively ended the existence of the Larsen B portion of the shelf.

Although these events received much attention and were thought to be symptomatic of global warming, the Ross Sea sector does not seem to be warming at present. It is likely that the emission of giant icebergs in this zone was an isolated event. Intense iceberg outbreaks, such as the one described above, may not necessarily be occurring with a greater frequency than in the past. Rather, they are more easily detected with the aid of satellites.

In the typically ice-free Southern Ocean, surveys of iceberg diameters show that most bergs have a typical diameter of 300–500 metres (1,000–1,600 feet), although a few exceed 1 km (0.6 mile). It is possible to calculate the flexural (bending) response of a tabular iceberg to long Southern Ocean swells, and it has been found that a serious storm is capable of breaking down most bergs larger than 1 km into fragments.

Arctic bergs are generally smaller than Antarctic bergs, especially when newly calved. The largest recorded Arctic iceberg (excluding ice islands) was observed off Baffin Island in 1882; it was 13 km (8 miles) long by 6 km (4 miles) wide and possessed a freeboard, or the height of the berg above the waterline, of 20 metres (65 feet). Most Arctic bergs are much smaller and have a typical diameter of 100–300 metres (330–1,000 feet). Owing to their origin in narrow, fast-flowing glaciers, many Arctic bergs calve into random shapes that often develop further as they fracture and capsize. Antarctic bergs also evolve by the erosion of the weak freeboard or via further calving into tilted shapes. Depending on the local shape of the ice shelf at calving, the surfaces of icebergs, even while still predominantly tabular, may be domed or concave.

Erosion and Melting

Most of the erosion taking place on Antarctic icebergs occurs after the bergs have emerged into the open Southern Ocean. Melt and percolation through the weak firn layer bring most of the freeboard volume to the melting point. This allows ocean wave action around the edges to penetrate the freeboard portion of the berg. Erosion occurs both mechanically and through the enhanced transport of heat from ocean turbulence. The result is a

tabular · wedge · drydock · dome · pinnacled · blocky

Icebergs are typically divided into six types. Encyclopædia Britannica, Inc.

wave cut that can penetrate for several metres into the berg. The snow and firn above it may collapse to create a growler (a floating block about the size of a grand piano) or a bergy bit (a larger block about the size of a small house). At the same time, the turbulence level is enhanced around existing irregularities such as cracks and crevasses. Waves eat their way into these features, causing cracks to grow into caves whose unsupported roofs may also collapse. Through these processes, the iceberg can evolve into a drydock or a pinnacled berg. (Both types are composed of apparently independent freeboard elements that are linked below the waterline.) Such a berg may look like a megalithic stone circle with shallow water in the centre.

In the case of Arctic icebergs, which often suffer from repeated capsizes, there is no special layer of weak material. Instead, the whole berg gradually melts at a rate dependent on the salinity (the salt concentration present in a volume of water) and temperature at various depths in the water column and on the velocity of the berg relative to the water near the surface.

On the basis of their observations of iceberg deterioration, American researchers W.F. Weeks and Malcolm Mellor have proposed a rough formula for predicting melt loss:

$$-Z = KD$$

where Z = loss in metres per day from the walls and bottom of the iceberg, K = a constant of order 0.12, and D = mean water temperature in °C averaged over the draft of the iceberg. This yields a loss of 120 metres (400 feet) from iceberg sides and bottom during 100 days of drift in water at 10 °C (50 °F)—a rate that corresponds quite well to survival times of icebergs in waters off the coast of Newfoundland as measured by the

International Ice Patrol. It has been suggested that if the melt rate could be reduced by interposing a layer of fabric between the ice and water, an iceberg could theoretically survive long enough to be towed across the Atlantic Ocean from Newfoundland to Spain for use as a water and power source.

In Arctic icebergs, erosion often leads to a loss of stability and capsizing. For an Antarctic tabular berg, complete capsize is uncommon, though tiltmeter measurements have shown that some long, narrow bergs may roll completely over a very long period. More common is a shift to a new position of stability, which creates a new waterline for wave erosion. When tabular icebergs finally fragment into smaller pieces, these smaller individual bergs melt faster, because a larger proportion of their surface relative to volume is exposed to the water.

THE DISTRIBUTION OF ICEBERGS AND THEIR DRIFT TRAJECTORIES

In the Antarctic, a freshly calved iceberg usually begins by moving westward in the Antarctic Coastal Current, with the coastline on its left. Since its trajectory is also turned to the left by the Coriolis force owing to Earth's rotation, it may run aground and remain stationary for years before moving on. For instance, a large iceberg called Trolltunga calved from the Fimbul Ice Shelf near the Greenwich meridian in 1967, and it became grounded in the southern Weddell Sea for five years before continuing its drift. If a berg can break away from the coastal current (as Trolltunga had done by late 1977), it enters the Antarctic Circumpolar Current, or West Wind Drift. This eastward-flowing system circles the globe at latitudes of 40°–60° S. Icebergs tend to enter this current system at four well-defined longitudes or "retroflection zones": the Weddell Sea, east of the Kerguelen Plateau at longitude 90° E, west of the Balleny Islands at longitude 150° E, and in the northeastern Ross Sea. These zones reflect the partial separation of the surface water south of the Antarctic Circumpolar Current into independently circulating gyres, and they imply that icebergs found at low latitudes may originate from specific sectors of the Antarctic coast.

Once in the Antarctic Circumpolar Current, the iceberg's track is generally eastward, driven by both the current and the wind. Also, the Coriolis force pushes the berg slightly northward. The berg will then move crabwise in a northeasterly direction so that it can end up at relatively low latitudes and in relatively warm waters before disintegrating. In November 2006, for instance, a chain of four icebergs was observed just off Dunedin (at latitude 46° S) on New Zealand's South Island. Under extreme conditions, such as its capture by a cold eddy, an iceberg may succeed in reaching extremely low latitudes. For example, clusters of bergs with about 30 metres (100 feet) of freeboard were sighted in the South Atlantic at 35°50′ S, 18°05′ E in 1828. In addition, icebergs have been responsible for the disappearance of innumerable ships off Cape Horn.

In the Arctic Ocean, the highest latitude sources of icebergs are Svalbard archipelago north of Norway and the islands of the Russian Arctic. The iceberg production from these sources is not large—an estimated 6.28 cubic km (1.5 cubic miles) per year in a total of 250–470 cubic km (60–110 cubic miles) for the entire Arctic region. An estimated 26 percent originates in Svalbard, 36 percent stems from Franz Josef Land, 32 percent is added by Novaya Zemlya, about 6 percent begins in Severnaya Zemlya, and 0.3 percent comes from Ushakov Island. Many icebergs from these sources move directly into the shallow Barents or Kara seas, where they run aground. Looping trails of

Satellite image of Scoresby Sund, Greenland. Jacques Descloitres, MODIS
Rapid Response Team, NASA/GSFC

broken pack ice are left as the bergs move past the obstacles. Other bergs pass through Fram Strait and into the East Greenland Current. As these icebergs pass down the eastern coast of Greenland, their numbers are augmented by others produced by tidewater glaciers, especially those from Scoresby Sund. Scoresby Sund is an inlet that is large enough to have an internal gyral circulation. Water driven by the East Greenland Current enters on the north side of the inlet and flows outward on the south side. This pattern encourages the flushing of icebergs from the fjord. In contrast, narrower fjords offer more opportunities for icebergs to run aground; they also experience an estuarine circulation pattern where outward flow at the surface is nearly balanced by an inward flow at depth. An iceberg feels both currents because of its draft and thus does not move seaward as readily as sea ice generated in the fjord.

As the increased flux of icebergs reaches Cape Farewell, most bergs turn into Baffin Bay, although a few "rogue" icebergs continue directly into the Labrador Sea, especially if influenced by prolonged storm activity. Icebergs entering Baffin Bay first move northward in the West Greenland Current and are strongly reinforced by icebergs from the prolific West Greenland glaciers. About 10,000 icebergs are produced in this region every year. Bergs then cross to the west side of the bay, where they move south in the Baffin Island Current toward Labrador. At the northern end of Baffin Bay, in Melville Bay, lies an especially fertile iceberg-producing glacier front produced by the Humboldt Glacier, the largest glacier in the Northern Hemisphere.

Some icebergs take only 8–15 months to move from Lancaster Sound to Davis Strait, but the total passage around Baffin Bay can take three years or more, owing to groundings and inhibited motion when icebergs are

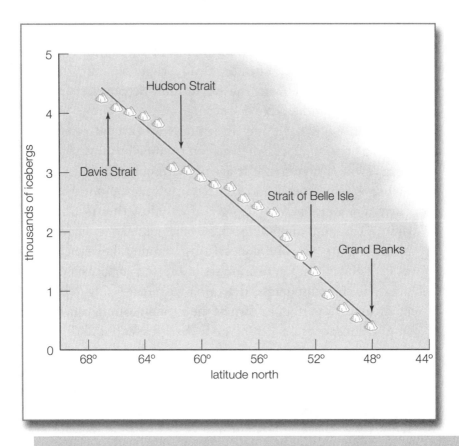

Graph of the change in iceberg number with decreasing latitude in the Northern Hemisphere. Encyclopædia Britannica, Inc.

embedded in winter sea ice. The flux of bergs that emerges from Davis Strait into the Labrador Current, where the final part of the bergs' life cycle occurs, is extremely variable. The number of bergs decreases linearly with latitude. This reduction is primarily due to melting and breakup or grounding followed by breakup. On average, 473 icebergs per year manage to cross the 48° N parallel and enter the zone where they are a danger to shipping—though numbers vary greatly from year to year. Surviving bergs will have lost at least 85 percent of their original mass. They are fated to melt on the Grand Banks or when they reach the "cold wall," or surface front, that separates the Labrador Current from the warm Gulf Stream between latitudes 40° and 44° N.

Much work has gone into modeling the patterns of iceberg drift, especially because of the need to divert icebergs away from oil rigs. It is often difficult to predict an iceberg's drift speed and direction, given the wind and current velocities. An iceberg is affected by the frictional drag of the wind on its smooth surfaces (skin friction drag) and upon its protuberances (form drag). Likewise, the drag of the current acts upon its immersed surfaces; however, the current changes direction with increasing depth, by means of an effect known as the Ekman spiral. Another important factor governing an iceberg's speed and direction is the Coriolis force, which diverts icebergs toward the right of their track in the Northern Hemisphere and toward the left in the Southern Hemisphere. This force is typically stronger on icebergs than on sea ice, because icebergs have a larger mass per unit of sea-surface area. As a result, it is unusual for icebergs to move in the same direction as sea ice. Typically, their direction of motion relative to the surface wind is some 40°–50° to the right (Northern Hemisphere) or left (Southern Hemisphere). Icebergs progress at about 3 percent of the wind speed.

ICEBERG SCOUR AND SEDIMENT TRANSPORT

When an iceberg runs aground, it can plow a furrow several metres deep in the seabed that may extend for tens of kilometres. Iceberg scour marks have been known from the Labrador Sea and Grand Banks since the early 1970s. In the Arctic, many marks are found at depths of more than 400 metres (1,300 feet), whereas the deepest known sill, or submerged ridge, within Greenland fjords is 220 metres (about 725 feet) deep. This unsolved anomaly suggests that icebergs were much deeper in the past or that sedimentation rates within the fjords are so slow that marks dating from periods of reduced sea level have not yet been filled in. It is also possible that an irregular berg can increase its draft by capsizing, though model studies suggest that the maximum gain is only a few percent. Since not all iceberg-producing fjords have been adequately surveyed, another possibility is that Greenland fjords exist with entrances of greater depth. In the Antarctic, the first scours were found in 1976 at latitude 16° W off the coast of Queen Maud Land in the eastern Weddell Sea, and further discoveries were made off Wilkes Land and Cape Hallett at the eastern entrance to the Ross Sea.

In addition, iceberg scour marks have been found on land. On King William Island in the Canadian Arctic, scour marks have been identified in locations where the island rose out of the sea—the result of a postglacial rebound after the weight of the Laurentide Ice Sheet was removed. Furthermore, Canadian geologist Christopher Woodworth-Lynas has found evidence of iceberg scour marks in the satellite imagery of Mars. Scour marks are strong indicators of past water flow.

Observations indicate that long furrows like plow marks are made when an iceberg is driven by sea ice, whereas a freely floating berg makes only a short scour mark or a single depression. Apart from simple furrows, "washboard patterns" have been seen. It is thought that these patterns are created when a tabular berg runs aground on a wide front and is then carried forward by tilting and plowing on successive tides. Circular depressions, thought to be made when an irregular iceberg touches bottom with a small "foot" and then swings to and fro in the current, have also been observed. Grounded bergs have a deleterious effect on the ecosystem of the seabed, often scraping it clear of all life.

Both icebergs and pack ice transport sediment in the form of pebbles, cobbles, boulders, finer material, and even plant and animal life thousands of kilometres from their source area. Arctic icebergs often carry a top burden of dirt from the eroded sides of the valley down which the parent glacier ran, whereas both Arctic and Antarctic bergs carry stones and dirt on their underside. Stones are lifted from the glacier bed and later deposited out at sea as the berg melts. The presence of ice-rafted debris (IRD) in seabed-sediment cores is an indicator that icebergs, sea ice, or both have occurred at that location during a known time interval. (The age of the deposit is indicated by the depth in the sediment at which the debris is found.)

Noting the locations of ice-rafted debris is a very useful method of mapping the distribution of icebergs and thus the cold surface water occurring during glacial periods and at other times in the geologic past. IRD mapping surveys have been completed for the North Atlantic, North Pacific, and Southern oceans. The type of rock in the debris can also be used to identify the source region of the transporting iceberg. Caution must be used in such interpretation because, even in the modern era, icebergs can spread far beyond their normal limits under exceptional conditions. For instance, reports of icebergs off the coast of Norway in spring 1881 coincided with the most extreme advance ever recorded of East Greenland sea ice. It is likely that the bergs were carried eastward along with the massive production and outflow of Arctic sea ice.

It is ice-rafted plant life that gives the occasional exotic colour to an iceberg. Bergs are usually white (the colour of

164

snow or bubbly ice) or blue (the colour of glacial ice that is relatively bubble-free). A few deep green icebergs are seen in the Antarctic; it is believed that these are formed when seawater rich in organic matter freezes onto the bottoms of the ice shelves.

THE CLIMATIC IMPACTS OF ICEBERGS

Most scientists maintain that adding large numbers of melting icebergs to the oceans may cause an increase in global sea level if the polar regions are not replenished by an equal amount of snowfall. Also, the addition of large amounts of freshwater to the oceans may lower the salinity of the upper layers of the ocean and possibly alter the present convection-current regime.

THE IMPACTS ON ICE SHEETS AND SEA LEVEL

Apart from local weather effects, such as fog production, icebergs have two main impacts on climate. Iceberg production affects the mass balance of the parent ice sheets, and melting icebergs influence both ocean structure and global sea level.

The Antarctic Ice Sheet has a volume of 28 million cubic km (about 6.7 million cubic miles), which represents 70 percent of the total fresh water (including groundwater) in the world. The mass of the ice sheet is kept in balance by a process of gain and loss—gain from snowfall over the whole ice sheet and ice loss from the melting of ice at the bottom of the ice shelf and from the calving of icebergs from the edges of the ice shelf. The effect of summer runoff and from sublimation off the ice surface is negligible.

Annual snowfall estimates for the Antarctic continent start at 1,000 cubic km (240 cubic miles). If the Antarctic

Ice Sheet is in neutral mass balance, the annual rate of loss from melting and iceberg calving must be close to this value; indeed, estimates of iceberg flux do start at this value, though some run much higher. Such apparently large fluxes are still less than the mean flow rate of the Amazon River, which is 5,700 cubic km (about 1,370 cubic miles) per year. In Antarctica the annual loss amounts to only one ten-thousandth of its mass, so the ice sheet is an enormous passive reservoir. However, if losses from iceberg calving and ice-shelf melting are greater than gains from snowfall, global sea levels will rise.

At present, the size, and even the sign, of the contribution from Antarctica is uncertain. Consequently, Antarctic ice flux has not been included as a term in the sea-level predictions of *Climate Change 2007*, the fourth assessment report of the Intergovernmental Panel on Climate Change (IPCC). What is more certain is that the retreat of glaciers in the Arctic and mountain regions has contributed about 50 percent to current rates of sea-level rise. (The rest is due to the thermal expansion of water as the ocean warms.) An increasing contribution is coming from a retreat of the Greenland Ice Sheet, and part of this contribution is occurring as an iceberg flux.

THE IMPACT ON OCEAN STRUCTURE

In considering the effect of iceberg melt upon ocean structure, it is found that the total Antarctic melt is equivalent to the addition of 0.1 metre (0.3 foot) of fresh water per year at the surface. This is like adding 0.1 metre of extra annual rainfall. The dilution that occurs, if averaged over a mixed layer 100–200 metres (330–660 feet) deep, amounts to a decrease of 0.015–0.03 part per thousand (ppt) of salt. Melting icebergs thus make a small but measurable contribution to maintaining the Southern Ocean pycnocline (the density boundary separating low-salinity surface water from higher-salinity deeper water) and to keeping surface salinity in the Southern Ocean to its observed low value of 34 ppt or below.

It is interesting to note that the annual production of Antarctic iceberg ice is about one-tenth of the annual production of Antarctic sea ice. Sea ice has a neutral effect on overall ocean salinity, because it returns to liquid during the summer months. Nevertheless, when sea ice forms, it has an important differential effect in that it increases ocean salinity where it forms. This is often near the Antarctic coast. Increased salinity encourages the development of convection currents and the formation of bottom water (masses of cold and dense water). Icebergs, on the other hand, always exert a stabilizing influence on the salinity of the water column. This stabilizing influence manifests itself only when the icebergs melt, and this occurs at lower latitudes.

Individual Arctic icebergs, although similar in numbers to Antarctic bergs (10,000–15,000 emitted per year), are smaller on average, so the ice flux is less. This, however, was not necessarily the case during the last glacial period. It has been postulated that, during the first stage of the retreat of the Laurentide Ice Sheet of North America, a large ice-dammed glacial lake (Lake Agassiz) formed in Canada over much of present-day Manitoba. When the ice dam broke, an armada of icebergs was suddenly released into the North Atlantic. As the icebergs melted, they added so much fresh water at the surface that the normal winter convection processes were turned off in the North Atlantic Ocean. As a result, the Gulf Stream was weakened, and northern Europe was returned to ice-age conditions for another millennium—the so-called Younger Dryas event.

ICEBERG DETECTION, TRACKING,
AND MANAGEMENT

An iceberg is a very large object that can be detected in the open sea both visually and by radar. In principle an iceberg can also be detected by sonar. In the open sea, an iceberg produces squealing, popping, and creaking sounds caused by mechanical stresses and cracking, and these sounds can be detected underwater up to 2 km (more than a mile) away. In summer, bergs can also produce a high-pitched hissing sound called "bergy seltzer," which is due to the release of high-pressure air bubbles from the ice as it melts in the warmer water.

The discovery of an iceberg depends on the alertness of a ship's watchkeepers. A decaying iceberg poses additional hazards because of its trail of growlers and bergy bits. Although small in size, they have masses (up to 120 tons for growlers, up to 5,400 tons for bergy bits) that are capable of damaging or sinking ships. As they drop into the sea, icebergs often roll over and lose their snow layers. In a heavy sea, the bergs' smooth wetted ice surfaces produce a low radar cross section. This makes them difficult to discriminate by eye against foam and whitecaps. Because a ship may steer to avoid a large parent berg, it may be in greater danger from undetected growlers or bergy bits drifting nearby.

The problem of protecting shipping from icebergs is most critical in two regions, the high-latitude Southern Ocean and the northwestern section of the North Atlantic. The Southern Ocean threat is increasing because large container ships—those unable or unwilling to use the Panama Canal—can reach high southern latitudes on transit from Australia or New Zealand to Cape Horn. No special measures are currently in place to protect such vessels. In the North Atlantic, the International Ice Patrol was established in 1914 following the loss of the RMS *Titanic* to an iceberg in April 1912. Its task is to track icebergs as they enter shipping lanes via the Labrador Current and to keep a continuous computer plot of the known or estimated whereabouts of every berg. Reports are transmitted twice a day to ships. In the past, iceberg positions were sited by ships or aircraft. However, it is becoming more common that icebergs are sited by the interpretation of satellite imagery.

The most useful type of sensor is synthetic aperture radar (SAR), which combines high resolution with day-and-night weather-independent capability. Tools with a pixel size of about 20 metres (65 feet) are capable of resolving most bergs. The new generation of SAR in the early 21st century, such as the Canadian RADARSAT and the European ENVISAT, also surveys wide swaths (up to 400 km [250 miles] wide) in every orbit and thus is capable of surveying the entire danger zone once per day.

During the 1950s and 1960s, attempts were made by the U.S. Coast Guard to find ways of fragmenting icebergs that posed a threat to shipping. All were unsuccessful. Explosive techniques were particularly so, since ice and snow are so effective at absorbing mechanical shock. Often the yield of fragmented ice was no greater than the mass of explosive used. Because of the need to defend offshore drilling and production platforms from icebergs, the viability of explosive techniques has been readdressed more recently. It was found that very cold ice, such as the type found in the lower part of an iceberg, can be fragmented successfully by the use of slow-burning explosives such as Thermit. Thermit can be implanted by drilling. However, implantation is a dangerous process because of the possibility of capsize.

Until these techniques are perfected, icebergs cannot be destroyed. Current protocols call for the location and tracking of threatening icebergs. Iceberg trajectories are then predicted by increasingly sophisticated computer models. If necessary, icebergs are captured and towed out of the way of their targets.

SEA ICE

Most sea ice occurs as pack ice, which is very mobile, drifting across the ocean surface under the influence of the wind and ocean currents and moving vertically under the influence of tides, waves, and swells. There is also landfast ice, or fast ice, which is immobile, since it is either attached directly to the coast or seafloor or locked in place between grounded icebergs. Fast ice grows in place by freezing of seawater or by pack ice becoming attached to the shore, seafloor, or icebergs. Fast ice moves up and down in response to tides, waves, and swells, and pieces may break off and become part of the pack ice. A third type of sea ice, known as marine ice, forms far below the ocean surface at the bottom of ice shelves in Antarctica. Occasionally seen in icebergs that calve from the ice shelves, marine ice can appear green due to organic matter in the ice.

Sea ice undergoes large seasonal changes in extent as the ocean freezes and the ice cover expands in the autumn and winter, followed by a period of melting and retreat in the spring and summer. Northern Hemisphere sea ice extent typically ranges from approximately 8 million square km in September to approximately 15 million square km in March. (One square km equals approximately 0.4 square mile.) Southern Hemisphere sea ice extent ranges from approximately 4 million square km in February to approximately 20 million square km in September. In

September 2007 the sea ice extent in the Northern Hemisphere declined to roughly 4.1 million square km, a figure some 50 percent below mean sea ice coverage for that time of year. Globally, the minimum and maximum sea ice extents are about 20 million square km and 30 million square km, respectively. Measured routinely using data obtained from orbiting satellite instruments, the minimum and maximum sea ice extent figures vary annually and by decade. These figures are important factors for understanding polar and global climatic variation and change.

ICE SALINITY, TEMPERATURE, AND ECOLOGICAL INTERACTIONS

As seawater freezes and ice forms, liquid brine and air are trapped within a matrix of pure ice crystals. Solid salt crystals subsequently precipitate in pockets of brine within the ice. The brine volume and chemical composition of the solid salts are temperature-dependent.

Liquid ocean water has an average salinity of 35 parts per thousand. New ice such as nilas has the highest average salinity (12–15 parts per thousand); as ice grows thicker during the course of the winter, the average salinity of the entire ice thickness decreases as brine is lost from the ice. Brine loss occurs by temperature-dependent brine pocket migration, brine expulsion, and, most importantly, by gravity drainage via a network of cells and channels. At the end of winter, Arctic first-year ice has an average salinity of 4–6 parts per thousand. Antarctic first-year ice is more saline, perhaps because ice growth rates are more rapid than in the Arctic, and granular ice traps more brine.

In summer, gravity drainage of brine increases as the ice temperature and permeability increase. In the Arctic, summer gravity drainage is enhanced by flushing, as snow and ice meltwater percolate into the ice. Consequently, after a few summers the ice at the surface is completely desalinated and the average salinity of Arctic multiyear ice drops to 3–4 parts per thousand. Antarctic multiyear ice is more saline because the snow rarely melts completely at the ice surface, and brine flushing is uncommon. Instead of percolating into the ice, snow meltwater refreezes onto the ice surface, forming a layer of hard, glassy ice. In contrast, even though it forms from platelets in seawater, marine ice contains little or no salt. The reasons for this remain unclear, but possible explanations include the densification of the ice crystals or their desalination by convection within the "mushy" crystal layer.

Because sea ice is porous and permeable and the brine held within it contains nutrients, sea ice often harbours rich and complex ecosystems. Viruses, bacteria, algae, fungi, and protozoans inhabit sea ice, taking advantage of the differences in salinity, temperature, and light levels. Algae are perhaps the most obvious manifestation of the sea ice ecosystem because they are pigmented and darken the ice. Algae are found at the top, bottom, and interior of Antarctic sea ice; however, they are found primarily at the bottom of Arctic sea ice, where they can occur as strands many metres in length. Sea ice algae are important as a concentrated food source for krill and other zooplankton. Melting sea ice rich in algae may also be important for seeding phytoplankton blooms in the previously ice-covered ocean.

SEA ICE FORMATION AND FEATURES

Sea ice that is not more than one winter old is known as first-year ice. Sea ice that survives one or more summers is known as multiyear ice. Most Antarctic sea ice is first-year pack ice. Multiyear ice is common in the Arctic, where most of it occurs as pack ice in the Arctic Ocean.

Pack ice is made up of many individual pieces of ice known as cakes, if they are less than 20 metres (about 66 feet) across, and floes, which vary from small (20–100 metres [about 66–330 feet] across) to giant (greater than 10 km [about 6 miles] across). As the ice drifts, it often breaks apart, and open water appears within leads and fractures. Leads are typically linear features that are widespread in the pack ice at any time of year, extend for hundreds of kilometres, and vary from a few metres to hundreds of metres in width. In winter, leads freeze quickly. Both new and young ice are often thickened mechanically by rafting and ridging, when they are compressed between thicker floes. A pressure ridge is composed of a sail above the waterline and a keel below. In the Arctic most keels are 10–25 metres (about 33–80 feet) deep and typically four times the sail height. Keel widths are typically 2–3 times the sail width. Antarctic pressure ridges are less massive than Arctic pressure ridges. Though they only make up about 25 percent of the total ice area in both polar regions, approximately 40–60 percent of the total ice mass is contained within pressure ridges.

Ice crystals growing on the ocean surface typically break down quickly into smaller pieces that form a soupy suspension known as frazil or grease ice. Under calm conditions the crystals freeze together to form a continuous sheet of new ice called nilas. It is up to 10 cm (about 4 inches) thick and looks dark gray. As the sheet ice thickens by freezing at the bottom, it becomes young ice that is gray to grayish white and up to 30 cm (about 1 foot) thick. If new and young ice are not deformed into rafts or ridges, they will continue to grow by a bottom-freezing process known as congelation. Congelation ice, with its

Pressure ridge in multiyear sea ice thrust up against the northernmost coast of Ellesmere Island, Queen Elizabeth Islands, Canada. M.O. Jeffries, University of Alaska Fairbanks

distinctive columnar crystal texture due to the downward growth of the ice crystals into the water, is very common in Arctic pack ice and fast ice.

Under more turbulent conditions, when the water is disturbed by wind and waves, frazil crystals agglomerate into discs known as pancakes. As they grow from a few centimetres to a few metres across, they solidify and thicken mechanically by rafting on top of each other. Pancakes freeze together to form cakes and floes, which contain a large amount of ice with a granular texture. The "frazil-pancake cycle," though it occurs in both hemispheres, is particularly important in Antarctica, where it accounts for the rapid expansion of ice cover during the autumn and winter. Consequently, Antarctic ice floes generally contain a larger amount of granular ice and a smaller amount of columnar ice than Arctic ice floes.

Frazil, grease, and pancake ice formation also occur in polynyas, which are recurrent features that remain partially or totally ice-free in areas normally expected to be covered with sea ice. They are particularly common in Antarctica, where katabatic winds blowing off the continent force the ice at the coast away from shore, leaving the ocean surface ice-free and open to further ice growth. Ice formation and removal can be almost continuous in coastal polynyas. Consequently, they are sometimes referred to as "ice factories."

Antarctic ice floes also contain a significant amount of granular ice because the weight of snow is often sufficient to depress the ice surface below sea level, soaking the base of the snow with seawater and producing a slush. When the slush freezes, a layer of granular snow ice is added near the top of the floes.

Platelet ice is perhaps the most exotic form of sea ice besides marine ice. In Antarctica, where cold, relatively low-salinity seawater flows out from beneath ice shelves, platelet ice grows both in the water column and at the bottom of the sea ice on the ocean surface. Whereas platelet ice has been found frozen into pack ice floes, it is most common in fast ice such as the type found in McMurdo Sound. In the Arctic, platelet ice grows primarily in pools of low-salinity water. These pools form at the base of ice floes during the summer months from meltwater runoff.

PACK ICE DRIFT AND THICKNESS

The large-scale drift of sea ice in the Arctic Ocean is dominated by the Beaufort Gyre (a roughly circular current flowing clockwise within the surface waters of the Beaufort Sea in the western or North American Arctic) and the Transpolar Drift (the major current flowing into the Atlantic Ocean from the eastern or Eurasian Arctic). The clockwise rotation of the Beaufort Gyre and the movement of the Transpolar Drift, the result of large-scale atmospheric circulation, are dominated by a high-pressure centre over the western Arctic Ocean. The pattern is not constant but varies in both strength and position about every decade or so, as the high-pressure centre weakens and moves closer to both Alaska and the Canadian Arctic. This decadal shift in the high-pressure centre is known as the Arctic Oscillation.

The Transpolar Drift exports large volumes of ice from the Arctic Ocean south through Fram Strait and along the east coast of Greenland into the North Atlantic Ocean. Ice drift speeds, determined from buoys placed on the ice, average 10–15 km (about 6–9 miles) per day in the Fram Strait. Ice can drift in the Beaufort Gyre for as much as seven years at rates that vary between zero at the centre to an average of 4–5 km (about 2.5–3 miles) per day at the edge. Together, the Beaufort Gyre and Transpolar

Drift strongly influence the Arctic Ocean ice thickness distribution, which has been determined largely from submarine sonar measurements of the ice draft. Ice draft is a measurement of the ice thickness below the waterline and often serves as a close proxy for total ice thickness. The average draft increases from about one metre (about three feet) near the Eurasian coast to 6–8 metres (about 20–26 feet) along the coasts of north Greenland and the Canadian Arctic islands, where the ice is heavily ridged.

In Antarctica the large-scale sea circulation is dominated by westward motion along the coast and eastward motion farther offshore in the West Wind Drift (also known as the Antarctic Circumpolar Current). The average drift speed is 20 km (about 12 miles) per day in the westward flow and 15 km (about 9 miles) per day in the eastward flow. Where katabatic winds force the ice away from the coast and create polynyas, local sea ice motion is roughly perpendicular to the shore. There are gyres in the Ross Sea and Weddell Sea where the westward-moving ice is deflected to the north and meets the eastward-moving ice further offshore. Unlike the Beaufort Gyre in the Arctic Ocean, these gyres do not appear to recirculate ice. Ice thickness data from drilling on floes, visual estimates by observers on ships, and a few moored sonars indicate that Antarctic sea ice is thinner than Arctic sea ice. Typically, Antarctic first-year ice is less than one metre (about three feet) thick, while multiyear ice is less than two metres (about 6.5 feet) thick.

SEA ICE AND ITS INTERACTIONS WITH THE OCEANS, ATMOSPHERE, AND CLIMATE

The growth and decay of sea ice influences local, regional, and global climate through interactions with the atmosphere and ocean. Whereas snow-covered sea ice is an

Frost smoke, open water, and new and young sea ice at a small lead surrounded by pack ice and icebergs in the Bellingshausen Sea, Antarctica. M.O. Jeffries, University of Alaska Fairbanks

effective insulator that restricts heat loss from the relatively warm ocean to the colder atmosphere, there is significant turbulent heat and mass transfer from leads and polynyas to the ocean and atmosphere during the winter months. These losses are manifested as frost smoke from evaporation and condensation at the water surface, and they affect atmospheric processes hundreds of metres above and hundreds of kilometres downstream from leads and polynyas. Brine rejected from ice growing within leads and polynyas drives the deep mixing of the ocean. Rejected brine also affects global ocean circulation and ventilation processes by increasing the salt concentration of the water it is released into. The conversion of both new and young ice into pressure ridges creates rough top and bottom surfaces that enhance the transfer of momentum from the atmosphere to the ocean. Ridges at the ice surface act as sails and catch the wind. The subsequent movement of the ice floes transfers energy to the underlying water via the keels on the underside of the ice.

Snow and ice reduce the amount of solar radiation available for organisms residing in the ice and water. This decrease in the amount of available energy affects and often reduces the productivity of plants, animals, and microorganisms. Snow has a high albedo (it reflects a significant proportion of solar shortwave radiation back to the atmosphere), and thus the temperature at the surface remains cool. In the Arctic the surface albedo decreases in summer as the snow melts completely, ponds of meltwater form on the ice surface that absorb a greater share of incoming shortwave radiation, and the overall ice concentration (the ratio of ice area to open water area) decreases. The increase in shortwave radiation absorption by meltwater ponds and the open ocean accelerates the melting process and further reduces surface albedo. This ice-albedo positive feedback plays a key role in the interaction of sea ice with climate.

THE EMERGING IMPACTS OF RECENT CHANGES TO SEA ICE

Submarine sonar data obtained since 1958 have revealed that the average ice draft in the Arctic Ocean in the 1990s decreased by over 1 metre (about 3 feet) and that ice volume was 40 percent lower than during the period 1958–76. The greatest ice draft reduction occurred in the central and eastern Arctic. Remote sensing also revealed a reduction of 3 percent per decade in Arctic sea ice extent from 1978, with particularly rapid losses occurring from the late 1980s. This included the eastern Arctic, where both the ice concentration and the duration of the ice-covered season also decreased. Computer simulations suggest that sea ice changes in this region were due to changes in atmospheric circulation, and thus ice dynamics, rather than higher air temperatures. Yet it is not clear whether these changes are due to natural variability—i.e., the Arctic Oscillation—or whether they represent a regime shift that will persist and perhaps become even more severe in the future.

Since computer models of climate change predict that the consequences of global warming will occur earlier and be most pronounced in the polar regions, particularly the Arctic, monitoring and understanding the behaviour of sea ice are important. Continued reductions of Arctic sea ice extent could have potentially severe ecological impacts. One such event may have arisen in western Hudson Bay, Canada, where a significant decline in the physical condition and reproductive success of polar bears occurred as the duration and extent of sea ice cover decreased during the 1980s and 1990s. On the other hand, a reduction in sea ice could be advantageous for oil and mineral exploration, production, and transport, and for navigation through the Northern Sea Route (Northeast Passage), a water route connecting the Atlantic and Pacific Oceans along the northern coast of Europe and Russia, and the Northwest Passage, a similar route along the northern coast of North America.

Whaling records suggest that Antarctic sea ice extent decreased by approximately 25 percent between the mid-1950s and early 1970s, whereas ice core samples suggest a 20 percent decrease in sea ice extent since 1950. Since then, remote sensing data have indicated an increase in Antarctic sea ice extent parallel to the decrease in Arctic sea ice extent through the 1980s and 1990s. Yet the increase in Antarctic sea ice extent has not been uniformly distributed. A reduction in sea ice extent west of the Antarctic Peninsula has been correlated with slight declines in Adélie penguin numbers and a significant rise in the Chinstrap penguin population. There is speculation that if ice extent continues to decrease in this region, krill numbers will diminish significantly as they lose their under-ice habitat and face growing competition from salps.

CHAPTER 7
THE ARCTIC AND ANTARCTIC

Earth's polar regions are immense storehouses of ice. The Arctic, made up of the Arctic Ocean and adjacent terrestrial regions, and Antarctica, the world's southernmost continent, are source regions for the world's cold air masses and cold ocean currents, providing a counterbalance to the warm air and warm water of the tropics.

THE ARCTIC

The Arctic is the northernmost region of the Earth, centred on the North Pole and characterized by distinctively polar conditions of climate, plant and animal life, and other physical features. The term is derived from the Greek *arktos* ("bear"), referring to the northern constellation of the Bear. It has sometimes been used to designate the area within the Arctic Circle—a mathematical line that is drawn at latitude 66°30′ N, marking the southern limit of the zone in which there is at least one annual period of 24 hours during which the sun does not set and one during which it does not rise. This line, however, is without value as a geographic boundary, since it is not keyed to the nature of the terrain.

While no dividing line is completely definitive, a generally useful guide is the irregular line marking the northernmost limit of the stands of trees. The regions north of the tree line include Greenland (Kalaallit Nunaat),

Svalbard, and other polar islands; the northern parts of the mainlands of Siberia, Alaska, and Canada; the coasts of Labrador; the north of Iceland; and a strip of the Arctic coast of Europe. The last-named area, however, is classified as subarctic because of other factors.

Conditions typical of Arctic lands are extreme fluctuations between summer and winter temperatures; permanent snow and ice in the high country; grasses, sedges, and low shrubs in the lowlands; and permanently frozen ground (permafrost), the surface layer of which is subject to summer thawing. Three-fifths of the Arctic terrain is outside the zones of permanent ice. The brevity of the Arctic summer is partly compensated by the long daily duration of summer sunshine.

International interest in the Arctic and subarctic regions has steadily increased during the 20th century, particularly since World War II. Three major factors are involved: the advantages of the North Pole route as a shortcut between important centres of population, the growing realization of economic potentialities such as mineral (especially petroleum) and forest resources and grazing areas, and the importance of the regions in the study of global meteorology.

Continental Ice Sheets of the Past

Little is known about the climate of the northern lands in early Cenozoic times. It is possible that the tree line was at least 1.6 km (1,000 miles) farther north than at present. During the Cenozoic, however, the polar lands became cooler and permanent land ice formed, first in the Alaskan mountain ranges and subsequently, by the end of the Pliocene (2.6 million years ago), in Greenland. By the onset of the Quaternary Period, glaciers were widespread

in northern latitudes. Throughout the Quaternary, conti-
nental-scale ice sheets expanded and decayed on at least
eight occasions in response to major climatic oscillations
in high latitudes. Detailed information available for the
final glaciation (80,000 to 10,000 years ago) indicates
that in North America the main ice sheet developed on
Baffin Island and swept south and west across Canada,
amalgamating with smaller glaciers to form the Laurentide
Ice Sheet, covering much of the continent between the
Atlantic Ocean and the Rocky Mountains and between
the Arctic Ocean and the Ohio and Missouri river valleys. A smaller ice cap formed in the Western Cordillera.
The northern margin of the ice lay along the Brooks
Range (excluding the Yukon Basin) and across the southern islands of the Canadian Archipelago. To the north
the Queen Elizabeth Islands supported small, probably
thin, ice caps. Glacier ice from Greenland crossed
Nares Strait to reach Ellesmere Island during maximum
glaciation.

The Atlantic Arctic islands were covered with ice
except where isolated mountain peaks (nunataks) projected through the ice. In Europe the Scandinavian
Ice Sheet covered most of northern Europe between
Severnaya Zemlya in Russia and the British Isles.
Northeastern Siberia escaped heavy glaciation, although,
as in northern Canada, the ice sheet had been more extensive in an earlier glaciation.

As the ice sheets melted, unique landforms developed
by the ice were revealed. Although not restricted to the
present Arctic, they are often prominent there and, in the
absence of forests, are clearly visible. In areas of crystalline rocks, including large parts of the northern Canadian
Shield and Finland, the ice left disarranged drainage and
innumerable lakes. In the lowlands deep glacial deposits
filled eroded surfaces and produced a smoother landscape,

often broken by low ridges and hills of glacial material, drumlins, rogen (ribbed) moraines, and eskers. In the uplands the characteristic glacial landforms are U-shaped valleys. Near the polar coasts these have been submerged to produce fjords, which are well developed in southern Alaska, along the east coast of Canada, around Greenland, in east and west Iceland, along the coast of Norway, and on many of the Arctic islands.

Because of their enormous weight, continental ice sheets depress the Earth's crust. As the ice sheets melted at the close of the Pleistocene Epoch (11,700 years ago), the land slowly recovered its former altitude, but before this was completed the sea flooded the coastal areas. Subsequent emergence has elevated marine beaches and

Polar bears walk along the rocky shores of Sillem Island in Canada. The weight of ice sheets and flooding over time has resulted in raised beaches in many Arctic locations. Pete Ryan/National Geographic/Getty Images

sediments to considerable heights in many parts of the Arctic, where their origin is easily recognized from the presence of marine shells, the skeletons of sea mammals, and driftwood. The highest strandlines are found 500 to 900 feet above contemporary sea level in many parts of the western and central Canadian Arctic and somewhat lower along the Baffin Bay and Labrador coasts. Comparable emergence is found on Svalbard, Greenland, the northern Urals, and on the Franz Josef Archipelago, where it reaches more than 1,500 feet. In many emerged lowlands, such as those south and west of Hudson Bay, the raised beaches are the most conspicuous features in the landscape, forming hundreds of low, dry, gravel ridges in the otherwise ill-drained plains. Emergence is still continuing, and in parts of northern Canada and northern Sweden uplift of two to three feet a century has occurred during the historical period. In contrast, a few Arctic coasts, notably around the Beaufort Sea, are experiencing submergence at the present time.

Polar continental shelves in areas that escaped glaciation during the ice ages were exposed during periods of low sea level, especially in the Bering Strait and Sea (Beringia), which facilitated migration of people to North America from Asia, and in the Laptev and East Siberian seas.

TERRAIN

Although the detail of the terrain in many parts of the Arctic is directly attributable to the Pleistocene glaciations, the major physiographic divisions reveal close correlation with geologic structure. The two largest shield areas, the Canadian and the Baltic, have developed similar landscapes. West of Hudson Bay, in southwestern Baffin Island, and in Karelia the land is low and rocky with countless lakes and disjointed drainage. Uplands, generally

1,000 to 2,000 feet above sea level and partially covered with glacial deposits, are more widely distributed. They form the interior of Quebec-Labrador and parts of the Northwest Territories in Canada, and the Lapland Plateau in northern Scandinavia. The eastern rim of the Canadian Shield in Canada from Labrador to Ellesmere Island has been raised by crustal changes and then dissected by glaciers to produce fjords that separate mountain peaks more than 6,000 feet high. The surface of the shield in Greenland has the shape of an elongated basin, with the central part, which is below sea level, buried beneath the Greenland ice cap. Around the margins, on the east and west coasts, the mountainous rim is penetrated by deep troughs through which local and inland-ice glaciers flow to the sea. The mountains are highest in the east, where they exceed 10,000 feet.

In shield areas where sedimentary rocks mantle the crystalline variety, as in north-central Siberia, the southern sector of the Canadian archipelago, and Peary Land, the topography varies from plains to plateaus, with the latter deeply dissected by narrow valleys. Far beyond the margins of the shields, extensive plains have evolved on soft sedimentary rocks. In North America these form the Mackenzie Lowlands, Banks and Prince Patrick islands, and the Arctic Plains section of northern Alaska; in northern Europe they form the Severnaya Dvina and Pechora Plains. In Siberia the Ob delta, its northeastern extension to the Laptev Sea, the North Siberian Lowland, the West Siberian Plain, and farther east the Lena-Kolyma plains (including the New Siberian Islands) have also developed on sedimentary rocks. Although there are differences in degree, these terrains are essentially flat, occasionally broken by low rock scarps, and covered with numerous shallow lakes. The plains are crossed by large rivers that have laid down deep alluvial deposits.

The strongly folded rocks associated with the two orogenic periods in the Arctic form separate physiographic regions. The original mountains of the older, Paleozoic folding were long ago destroyed by erosion, but the rocks have been elevated in recent geologic time, and renewed erosion, often by ice, has produced a landscape of plateaus, hills, and mountains very similar to the higher parts of the shields. In Ellesmere Island the mountains are nearly 10,000 feet high. In Peary Land and Spitsbergen, maximum elevations are about 6,000 feet, while in eastern Svalbard and on Novaya Zemlya and Severnaya Zemlya the uplands rarely exceed 2,000 feet. The younger groups of fold mountains of northeast Siberia and Alaska are generally higher. Peaks of 10,000 feet are found in the Chersky Mountains, 15,000 feet in Kamchatka, and even higher in southern Alaska. Characteristic of this physiographic division are wide intermontane basins drained by large rivers, including the Yukon and Kolyma.

Throughout the Arctic, excluding a few maritime areas, the winter cold is so intense that the ground remains permanently frozen except for a shallow upper zone, called the active layer, which thaws during the brief summer. Permanently frozen ground (permafrost) covers nearly one-quarter of the Earth's surface. In northern Alaska and Canada scattered observations suggest that permafrost is 800 to 1,500 feet deep; it is generally deeper in northern Siberia. The deepest known permafrost is in northern Siberia, where it exceeds 2,000 feet. The depth of the permafrost depends on the site, climate, vegetation, and recent history of the area, particularly whether it was covered by sea or glacier ice. Very deep permafrost was probably formed in unglaciated areas during the extreme cold of the ice ages. To the south in the subarctic, the permafrost thins and eventually becomes discontinuous, although locally it may still be 200 to 400 feet thick; along its southern boundary, permafrost survives under peat and in muskeg. In areas of continuous permafrost the active layer may be many feet thick in sandy well-drained soils with little vegetation but is usually less than six inches thick beneath peat.

Permafrost occurs in both bedrock and surface deposits. It has little effect in most rocks, but in fine-grained, unconsolidated sediments, particularly silts, lenses of ice, called ground ice, grow by migration of moisture, and in extreme cases half the volume of Arctic silts may be ice. Ground ice is often exposed in riverbanks and sea cliffs, where it may be 6 to 9 metres (20 to 30 feet) thick. In northern Siberia fossil ice has been reported up to 60 metres (200 feet) thick, although it may be glacier or lake ice that has subsequently been buried under river deposits. If ground ice melts, owing to a change in climate, hollows develop on the surface and quickly fill with water to form lakes and ponds. When frozen the silts have considerable strength, but if they thaw they change in volume, lose their strength, and may turn to mud. Variations in volume and bearing capacity of the ground due to changes in the permafrost constitute one of the major problems in Arctic construction.

DRAINAGE AND SOILS

Continuous permafrost inhibits underground drainage. Consequently, shallow lakes are numerous over large areas of the Arctic, and everywhere in early summer there is a wet period before the saturated upper layers of the ground dry out. During the summer waterlogged active layers on slopes may flow downhill over the frozen ground, a phenomenon known as solifluction. It is ubiquitous in the Arctic but is particularly intense where the soils are fine-grained, as in the coastal plain of northern Alaska, or where the precipitation is heavy, as on Bear Island in the

Norwegian Sea. The effect of solifluction is to grade slopes so that long, smooth profiles are common; slopes are normally covered with vegetation, but if the soil movement is too rapid plants may not be able to survive. Under these conditions the surface material is often graded, with narrow strips of pebbles and boulders separated by broader strips of finer particles.

The surface of many soils in northern areas show distinctive patterns produced by complex processes of freezing and thawing, which cause frost heaving and sorting of debris; although permafrost is not essential to these formations, it is usually present. There are many different types of patterned ground. In some, coarser material, pebbles, and boulders form polygonal nets, with the finer materials concentrated in the centre. When sorting is widely spaced, stone circles develop. Another variety of pattern, formed in sands and muds, is outlined by frost-crack fissures or strips of vegetation. Individual polygons vary from about half a metre (1.6 feet) to more than 90 metres (300 feet) in diameter. Mounds due to frost heaving in the soil also are widespread. They grow rapidly, disrupting leveled fields in a few years and limiting the use of farm machinery for haying. Elsewhere, notably in the Mackenzie valley and in parts of Alaska, removal of the natural vegetation—and, in isolated cases, plowing—has modified the soil climate. The ground ice has thawed, leading to disruption of drainage. Where the ice was wedge-shaped and in polygonal patterns, soil mounds several feet high may result. All Arctic terrains are sensitive to human-induced thermal disturbance, especially by vehicular traffic or oil-pipeline operations, and the preservation of the original soil climate is of great environmental importance.

The largest ice-covered mounds, which may reach 60 metres (200 feet) in height, are known in North America as pingos. Although they are widely distributed in the

Arctic and subarctic, major concentrations are restricted to the Mackenzie delta, the Arctic slope of Alaska, and coastal areas near the deltas of the Ob, Lena, and Indigirka rivers. Submarine landforms resembling pingos are found beneath the Beaufort Sea.

Arctic soils are closely related to vegetation. Unlike soils farther south, they rarely develop strong zonal characteristics. By far the most common are the tundra soils, which are circumpolar in distribution. They are badly drained and strongly acid and have a variable, undecomposed organic layer over mineral horizons. Some of the drier heath and grassland tundras overlie Arctic brown soils, which have a dark-brown upper horizon with gray and yellowish brown lower horizons. The active layer in the permafrost is normally deep in them.

Many exposed rock surfaces in the Arctic have been broken up by frost action so that the bedrock is buried under a cover of angular shattered boulders. These mantles are known as *felsenmeer* (German: "sea of rock") and are found principally on Arctic uplands. Their continuity and depth varies with climate, vegetation, and rock type, but they may be as much as 4 metres (12 feet) deep. Felsenmeer are especially well-developed on basalts and are consequently numerous on the basaltic Icelandic plateaus. They also develop quickly on sedimentary rocks and are wides-pread in the Canadian Arctic, where they occur down to sea level.

PRESENT-DAY GLACIATION

Although the Arctic is commonly thought to be largely ice-covered, less than two-fifths of its land surface in fact supports permanent ice. The remainder is ice-free because of either relatively warm temperatures or scant snowfall. Glaciers are formed when the annual accumulation of snow, rime, and other forms of solid precipitation exceeds that removed by summer melting. The excess snow is converted slowly into glacier ice, the rate depending on the temperature and annual accumulation of snow. In the Arctic, where most glaciers have temperatures far below the freezing point, the snow changes into ice slowly. In northwestern Greenland a hole 427 metres (1,400 feet) deep was drilled into the ice sheet without reaching glacier ice. The hole showed more than 800 annual snow layers, from which it was possible to determine precipitation changes for the past eight centuries. An ice core 1,400 metres (4,560 feet) deep was recovered in the mid-1960s from Camp Century in northwestern Greenland, and a core 2,000 metres (6,683 feet) deep from Dye 3, southeastern Greenland, was recovered in 1981. The ice cores have been analyzed for paleoclimatic and paleoatmospheric information covering the 100,000 years since the last interglacial.

The elevation at which accumulation and melting of glacier ice are equal is known as the equilibrium line and is roughly equivalent to the snow line. It frequently varies greatly over short distances and from year to year on a specific glacier. On Baffin Island the equilibrium line is a little more than 2,000 feet above sea level in the extreme southeast, rising to more than 4,500 feet in the Penny Ice Cap 300 miles to the north and descending to about 2,000 feet in the north of the island. In Greenland the line is at about 6,000 feet in the south and decreases irregularly to about 3,300 feet in the north. The summits of some ice caps are well below the snow line, but they continue to survive because of their low internal temperatures; the winter snowfall melts completely but refreezes in contact with the cold ice before flowing off the glacier. This phenomenon, first observed on the Barnes Ice Cap of Baffin Island, is now known to be widespread in the high Arctic.

GLACIER GROUPS

In Arctic Canada glacier ice is restricted, with few exceptions, to the northeast as a consequence of the greater
relief and precipitation around Baffin Bay and Davis Strait.
The most southerly ice is found in the Torngat Mountains
of northern Labrador, where there are small cirque glaciers
at the base of the mountains. Immediately north of Hudson
Strait on the plateau south of Frobisher Bay, there are two
small ice caps. Larger ice caps and highland ice (through
which mountains project) are present farther north along
the east of Baffin Island and on Bylot Island; only the
Barnes Ice Cap lies west of the coastal group. North of
Lancaster Sound the ice is more extensive, and large parts
of Devon, Ellesmere, and Axel Heiberg islands are glacier-
ized. In many ways these ice caps are small versions of the
Greenland Inland Ice, with a central dome-shaped section
and outlet glaciers flowing through the mountains toward
the sea. The ice cap on Meighen Island, the most westerly
of the group, is an exception, as it is circular in shape and
lies on low ground. Except for three small glaciers on
Melville Island, there are no glaciers in the Canadian western Arctic. Few Canadian glaciers reach the sea and form
icebergs. In the Arctic Ocean off northwestern Ellesmere
Island there is an area of floating shelf ice that may at one
time have been joined by glaciers, but the glaciers no longer reach the sea. This shelf ice has been the principal
source of the ice islands of the Arctic Ocean.

Other glaciers are found north and east of the Atlantic
Ocean and its continuation in the Norwegian and Barents
seas. Iceland has five major ice caps, the largest of which,
Vatna Glacier, covers more than 4,800 sq km (3,000 square
miles). All have small outlet glaciers, although none reaches
the sea. The ice caps owe their survival to heavy snowfall.

A grouping of glaciers on Spitsbergen, the largest island in Norway's Svalbard archipelago. The abundance of glaciers in the area make it a prime ecotourism destination. Chris Jackson/Getty Images

The western part of Vatna Glacier buries a volcano, Grímsvötn (Gríms Depression), which erupts every 6 to 10 years; the heat of the eruption forms a subglacial lake that bursts in great floods over the margins of the glacier.

North of Iceland, Jan Mayen Island supports a glacier on the volcano Mount Beeren. The glaciers of Svalbard cover about 90 percent of the land. On the largest island, Spitsbergen, the plateaus are covered with highland ice from which outlet glaciers reach the sea; there are also numerous independent valley and cirque glaciers. North East Land, the second largest island, supports two ice caps on its plateaus. On the east side of the Norwegian Sea, precipitation is heavy over the Scandinavian highlands, but temperatures are also high, and the total area of ice is only about 3,200 sq km (2,000 square miles), a small part of which is in northern Sweden and the remainder in Norway. To the northeast beyond the Barents Sea, precipitation is less, but the summer is shorter and permanent ice is widespread.

Farthest north in this group are the islands of the Franz Josef archipelago. Although at no point are they higher than 762 metres (2,500 feet), probably more than 90 percent of their area is covered with ice; some of the smaller islands are completely buried by glaciers. The southern island of Novaya Zemlya supports a few small glaciers; on the northern island they are more numerous, and the northern four-fifths of the island is ice-covered, with large outlet glaciers reaching the sea. Cyclonic depressions penetrate from the Barents Sea into the Kara Sea beyond Novaya Zemlya and produce sufficient snow for glaciers to form on Severnaya Zemlya. There are four major and many minor islands in the group. Although they are low-lying, consisting primarily of plateaus less than 610 metres (2,000 feet) high, all the larger islands have ice caps that cover less than half the total area. Outlet glaciers reach the sea and are an occasional source of icebergs. Elsewhere the Russian northern areas are remarkably free of glacier ice. Small cirque glaciers are found in the northern Ural Mountains and the Byrranga Mountains of the Taymyr Peninsula.

The glaciers around the North Pacific are concentrated in Alaska. The glaciers of southern Alaska are Alpine rather than Arctic and include some of the most spectacular mountain glaciers in the world. All types of ice are present, from small valley glaciers to highland ice that almost buries mountain ranges, with piedmont glaciers spreading out in the lowlands. The largest ice fields are around the Fairweather Range, the St. Elias Mountains, and the Chugach Mountains. Glaciers in these areas include the Hubbard, 145 km (90 miles) long, intermontane glaciers such as the Seward, and piedmont glaciers such as the Malaspina. Smaller glaciers also occur inland on the Alaska Range and in the Brooks Range of northern Alaska. There is more ice farther east in the Romanzof Mountains, where one glacier, the Okailak, is 10 miles long, and in a similar situation in the Selwyn and Ogilvie mountains of Canada's Yukon. There are a few small glaciers in the Aleutian Range and on the Aleutian Islands. On the northwest side of the Pacific basin there are small glaciers in the East Siberian Mountains and on the volcanic peaks of the Kamchatka Peninsula.

The overwhelming majority of Arctic glaciers for which precise data are available have experienced negative mass balances (i.e., reduction in mass) in the 20th century broken only by temporary cool phases in the 1960s and 1970s. The effect has been a general retreat of glacier fronts and thinning of ice around the margins. The Greenland Inland Ice may be an important exception to this generalization.

In Iceland, where glacier fluctuations are well recorded, the ice appears to have been restricted from the 10th until about the 16th century. The ice then advanced, reaching a maximum about 1750. A second advance followed a minor retreat, culminating about 1850, and a major retreat set in about 1890. The recession was slow at first, but by the 1930s it was generally rapid and has continued since, except locally for a brief interruption in the 1970s.

CLIMATE

The climates of polar lands vary greatly depending on their latitude, proximity of the sea, elevation, and topography; even so, they all share certain "polar" characteristics. Owing to the high latitudes, solar energy is limited to the summer months. Although it may be considerable, its effectiveness in raising surface temperatures is restricted by the high reflectivity of snow and ice. Only in the central polar basin does the annual net radiation fall below zero. In winter, radiative cooling at the surface is associated with extreme cold, but, at heights a few thousand feet above the surface, temperatures as much as 20 to 30°F (11 to 17°C) warmer can often be found. Temperature inversions such as this occur more than 90 percent of the time in midwinter in northwestern Siberia and over much of the polar basin. They also are common over the Greenland Ice Cap and in the sheltered mountain valleys of the Yukon and Yakutia. The lowest surface temperature ever recorded in North America was observed at Snag, Yukon (-81°F, -63°C), and even lower temperatures have been observed in Yakutia (-90°F, -68°C) and northern Greenland (-94°F, -70°C).

It has been customary to divide polar climates into two large groups, those corresponding to the climate of ice caps, in which no mean monthly temperature exceeds 32°F (0°C), and the tundra climates, with at least one month above 32°F but no month above 50°F (10°C). A more satisfactory division is to classify them as polar maritime climates, located principally on the northern islands and the adjacent coasts of the Atlantic and Pacific oceans, in which winter temperatures are rarely extremely low and snowfall is high; and the polar continental climates, as in northern Alaska, Canada, and Siberia, where winters are intensely cold and snowfall is generally light. Included in the polar continental climate type are the islands of the Canadian Arctic Archipelago, which are influenced only slightly by the sea in winter because of thick, unbroken sea ice. In addition to these two climates, there are smaller transitional zones, limited areas of "ice" climates, the climate of the polar basin, and, on the south side of the tree line, the subarctic climates.

In the polar continental areas, winter sets in toward the end of August in the far north and about a month later nearer the tree line. Temperatures continue to drop rapidly until about December. January, February, and early March have uniform conditions with mean temperatures about -35 °F (-37 °C) in the central Siberian Arctic and -30 to -20 °F (-34 to -29 °C) in North America. The lowest extreme temperatures in the winter are between -65 and -50 °F (-54 and -46 °C). A better indication of low temperatures as they affect humans is given by the windchill, a measurement of the cooling power of the atmosphere on human skin. It reaches a maximum north of Hudson Bay, where strong and persistent northwest winds, typical of the Canadian eastern Arctic, are combined with low air temperatures. This area is stormy in winter, with moderately high snowfall (1,300 to 2,500 mm [50 to 100 inches]), rapidly changing temperatures, and even occasional rain.

Elsewhere the winter continental climate is quiet, with long periods of clear sky and low snowfall. Visibility may be poor locally if there are open channels of water in the sea ice, and it is universally reduced when the wind blows drifting snow. The lowest snowfall is in the polar deserts of the northern Canadian islands and northern Greenland, where the total annual precipitation is frequently less than the equivalent of four inches of water.

Winter in the maritime Arctic (the Aleutians, coastal southwestern Greenland, Iceland, and the European Arctic) is a period of storminess, high winds, heavy precipitation in the form of either snow or rain (the latter at sea level), and moderate temperatures. The mean temperature of the coldest month is rarely below 20 °F (-7 °C), and extremely low temperatures are unknown.

Summer temperatures are more uniform across the whole of the Arctic. On the southern margin the monthly mean temperature reaches 50 °F (10 °C), and in continental situations short spells of hot weather with temperatures in the 80s°F (27–32°C), continuous sunshine, and calm weather are not uncommon; such weather often ends with thunderstorms. In the maritime climates, along the coasts, and on the northern islands when there is open water in the sea ice, the summer is relatively cool. In the south the temperatures are about 45°F (7°C), decreasing north to 40°F (4°C) or less; a maximum of 60°F (16°C) is hardly ever reached except at the heads of fjords as in southwestern Greenland, where marine influences are less marked. Fog and low clouds are widespread in maritime areas, and at this time of the year these areas are the cloudiest in the world. In lands that experience continental winters, precipitation is heaviest during the summer months; light rain and snow showers are frequent, but the average fall is low. The summer is everywhere a time of sudden changes. Calm, clear weather with sunshine and temperatures of about 50°F (10°C) will be followed by sudden winds, often causing a temperature drop of 20 to 30°F (11 to 17°C) and accompanied by cloud and fog.

Frost-free and growing periods are relatively short throughout the Arctic. For the most part there is no true frost-free period; frost and some snow have been recorded in every month of the year. At a few places near the tree line, notably in the Canadian western Arctic, the frost-free period may be the same as the less favourable parts of the prairies.

South of the tree line in the subarctic, differences between continental (Mackenzie Basin, interior Yukon, and Alaska and northeastern Siberia) and oceanic (northern Quebec-Labrador, northern Scandinavia, and northern Russia) situations are marked. A summer maximum of precipitation and frequent high summer temperatures (July means exceeding 60°F [16°C] in northeastern Siberia) in the continental regions contrast with heavier

precipitation, often with a fall maximum, and lower summer temperatures in the oceanic regions.

The central polar ocean, together with the Beaufort
and East Siberian seas, have winters comparable to northern Alaska and northeastern Siberia. Conditions are stable
for extended periods of low wind velocities, clear skies—
especially bordering Siberia—and temperatures ranging
from -20 to -40°F (-30 to -40°C). Occasional storms originating in the Barents and Bering seas may penetrate the
adjacent sectors of the polar basin and bring a temporary
rise in temperature accompanied by snow or blowing
snow. There is a negligible area (less than 1 percent) of
open water in the central polar basin in winter. By April,
air temperatures are rising until in June melting of the
snow and underlying sea ice begins. Mean summer temperatures fail to rise above 34°F (1°C) and are accompanied
by almost continuous low cloud cover and fog.

The only extensive ice climate in the Northern
Hemisphere is in Greenland. In the south the climate of
the inland ice cap has maritime characteristics with heavy
precipitation, mainly snow from passing cyclone distur-
bances. In the centre and north a continental situation
develops, and the snowfall is less. Although the air temperature may sometimes rise to 32°F (0°C), the mean
temperature is much lower than in the south. Strong
winds blowing off the ice cap are common in all parts of
the island.

The evidence from glacier fluctuations suggests significant climatic change in polar latitudes in the past
millennium. The first half of the 20th century saw climatic
amelioration in the Arctic, with higher temperatures
found particularly in winter and especially around the
Norwegian Sea. In general, the magnitude of the warming increased with latitude, and in Svalbard winter
temperatures rose by 14°F (8°C). Associated with climatic

changes were a radical reduction of sea ice around Svalbard and off southwestern Greenland. Birds, animals, and especially fish appeared farther north than before; in Greenland this led to a change in the economy, as its traditional dependence on seals yielded to dependence on fishing, particularly cod, which were caught north of the 70th parallel.

In the early 1940s, however, there was a downturn in polar temperatures. This widespread climatic cooling continued intermittently into the early 1970s. At this time sea ice failed to leave coastal areas in the summer in the eastern Canadian Arctic for the first time in living memory. A reversal of this trend followed in the next two decades, with the most noticeable temperature increases occurring in the lands to the north of the Pacific Ocean and around the Barents and Greenland seas (a change of +2.7 °F [+1.5 °C] in annual temperatures).

The underlying cause of the changes is not known, although they result directly from increased penetration of southerly winds into the polar regions.

THE ARCTIC OCEAN

Smallest of the world's oceans, centring approximately on the North Pole, the Arctic Ocean and its marginal seas (the Chukchi, East Siberian, Laptev, Kara, Barents, White, Greenland, and Beaufort; some oceanographers also include the Bering and Norwegian seas) are the least-known basins and bodies of water in the world ocean. This lack of knowledge about them results from their remoteness, hostile weather, and perennial or seasonal ice cover. This is changing, however, because the Arctic may exhibit a strong response to global change and may be capable of initiating dramatic climatic changes through alterations induced in the oceanic thermohaline circulation by its cold, southward-moving currents or through its effects on the global albedo resulting from changes in its total ice cover.

Although the Arctic Ocean is by far the smallest of the Earth's oceans, having only a little more than one-sixth the area of the next largest, the Indian Ocean, its area of 14,090,000 square km (5,440,000 square miles) is five times larger than that of the largest sea, the Mediterranean. The deepest sounding obtained in Arctic waters is 5,502 metres (18,050 feet), but the average depth is only 987 metres (3,240 feet).

Distinguished by several unique features, including a cover of perennial ice and almost complete encirclement by the landmasses of North America, Eurasia, and Greenland, the north polar region has been a subject of speculation since the earliest concepts of a spherical Earth. From astronomical observations, the Greeks theorized that north of the Arctic Circle there must be a midnight sun at midsummer and continual darkness at midwinter. The enlightened view was that both the northern and southern polar regions were uninhabitable frozen wastes, whereas the more popular belief was that there was a halcyon land beyond the north wind where the sun always shone and people called Hyperboreans led a peaceful life. Such speculations provided incentives for adventurous men to risk the hazards of severe climate and fear of the unknown to further geographic knowledge and national and personal prosperity.

OCEANOGRAPHY

Several factors in the Arctic Ocean make its physical, chemical, and biological processes significantly different from those in the adjoining North Atlantic and Pacific Oceans. Most notable is the covering ice pack, which reduces the exchange of energy between ocean and atmosphere by about 100 times. In addition, sea ice greatly reduces the penetration of sunlight needed for the photosynthetic processes of marine life and impedes the mixing effect of the winds. A further significant distinguishing feature is the high ratio of freely connected shallow seas to deep basins. Whereas the continental shelf on the North American side of the Arctic Ocean is of a normal width (approximately 40 miles), the Eurasian sector is hundreds of miles broad, with peninsulas and islands dividing it into five main marginal seas: the Chukchi, East Siberian, Laptev, Kara, and Barents. These marginal seas occupy 36 percent of the area of the Arctic Ocean, yet they contain only 2 percent of its water volume. With the exception of the Mackenzie River of Canada and the Colville River of Alaska, all major rivers discharge into these marginal shallow seas. The combination of large marginal seas, with a high ratio of exposed surface to total volume, plus large summer inputs of fresh water, greatly influences surface-water conditions in the Arctic Ocean.

As an approximation, the Arctic Ocean may be regarded as an estuary of the Atlantic Ocean. The major circulation into and from the Arctic Basin is through a single deep channel, the Fram Strait, which lies between the island of Spitsbergen and Greenland. A substantially smaller quantity (approximately one-quarter of the volume) of water is transported southward through the Barents and Kara seas and the Canadian Archipelago. The combined outflow to the Atlantic appears to be of major significance to the large-scale thermohaline circulation and mean temperature of the world ocean with a potentially profound impact on global climate variability. Warm waters entering the Greenland/Iceland/Norwegian (GIN) Sea plunge downward when they meet the colder waters from more northerly produced freshwater, southward-drifting ice, and a colder atmosphere. This produces North Atlantic Deep Water (NADW), which circulates in the world ocean. An increase in this fresh-water and ice export could shut down the thermocline convection in the GIN Sea; alternatively, a decrease in ice export might allow for convection and ventilation in the Arctic Ocean itself.

Low-salinity waters enter the Arctic Ocean from the Pacific through the shallow Bering Strait. Although the mean inflow seems to be driven by a slight difference in sea level between the North Pacific and Arctic oceans, a large source of variability is induced by the wind field, primarily large-scale atmospheric circulation over the North Pacific. The amount of freshwater entering the Arctic Ocean is about 2 percent of the total input. Precipitation is believed to be about 10 times greater than loss by evaporation, although both figures can be only roughly estimated. Through all these various routes and mechanisms, the exchange rate of the Arctic Ocean is estimated to be approximately 5.9 million cubic metres (210 million cubic feet) per second.

All waters of the Arctic Ocean are cold. Variations in density are thus mainly determined by changes in salinity. Arctic waters have a two-layer system: a thin and less dense surface layer is separated by a strong density gradi-ent, referred to as a pycnocline, from the main body of water, which is of quite uniform density. This pycnocline restricts convective motion and the vertical transfer of heat and salt, and hence the surface layer acts as a cap over the larger masses of warmer water below.

Despite this overall similarity in gross oceanographic structure, the waters of the Arctic Ocean can be classified into three major masses and one lesser mass.

1. The water extending from the surface to a depth of about 200 metres (about 650 feet) is the most variable and

heterogeneous of all that in the Arctic. This is because of the latent heat of freezing and thawing; brine addition from the process of ice freezing; freshwater addition by rivers, ice melting, and precipitation; and great variations in insolation (rate of delivery of solar energy) and energy flux as a result of sea ice cover. Water temperature may vary over a range of 7°F (4°C) and salinity from 28 to 34 grams of salt per kilogram of seawater (28 to 34 parts per thousand [°/oo]).

2. Warmer Atlantic water everywhere underlies Arctic surface water from a depth of about 200 to 914 metres (650 to 3,000 feet). As it cools it becomes so dense that it slips below the surface layer on entering the Arctic Basin. The temperature of this water is about 34 to 3 °F (1 to 3°C) as it enters the basin, but it is gradually cooled so that by the time it spreads to the Beaufort Sea it has a maximum temperature of 32.9 to 33.1°F (0.5 to 0.6°C). The salinity of the Atlantic layer varies between 34.5 and 35°/oo.

3. Bottom water extends beneath the Atlantic layer to the ocean floor. This is colder than the Atlantic water (below 32°F, or 0°C) but has the same salinity.

4. An inflow of Pacific water can be observed in the Amerasia Basin but not in the Eurasia Basin. This warmer and fresher water mixes with colder and more saline water in the Chukchi Sea, where its density enables it to flow as a wedge between the Arctic and Atlantic waters. The Pacific water, by the time it reaches the Canada Basin, has a temperature range of 31.1 to 30.8°F (-0.5 to -0.7°C) and salinities between 31.5 and 33°/oo.

Arctic waters are driven by the wind and by density differences. The net effect of tides is unknown but could have some modifying effect on gross circulation. The motion of surface waters is best known from observations of ice drift. The most striking feature of the surface circulation pattern is the large clockwise gyre (circular motion) that covers almost the entire Amerasia Basin. Fletcher's Ice Island (T-3) made two orbits in this gyre over a 20-year period, which is some indication of the current speed. The northern extremity of the gyre bifurcates and jets out of the Greenland-Spitsbergen passage as the East Greenland Current, attaining speeds of 15 to 40 centimeters (6 to 16 inches) per second. Circulation of the shallow Eurasian shelf seas seems to be a complex series of counterclockwise gyres, complicated by islands and other topographic relief.

Circulation of the deeper Atlantic water is less well known. On entering the Eurasia Basin, the plunging Greenland Sea water appears to flow eastward along the edge of the continental margin until it fans out and enters the Amerasia Basin along a broad front over the crest of the Lomonosov Ridge. There seems to be a general counterclockwise circulation in the Eurasia Basin and a smaller clockwise gyre in the Beaufort Sea. Speeds are slow—probably less than two inches per second.

The circulation of the bottom water is unknown but can be inferred to be similar to the Atlantic layer. Measured values of dissolved oxygen show that the bottom water is well ventilated, dissolved oxygen everywhere exceeding 70 percent of saturation.

SEA ICE IN THE ARCTIC OCEAN

The cover of sea ice suppresses wind stress and wind mixing, reflects a large proportion of incoming solar radiation, imposes an upper limit on the surface temperature, and impedes evaporation. Wind and water stresses keep the ice pack in almost continuous motion, causing the formation of cracks (leads), open ponds (polynya), and pressure ridges. Along these ridges the pack ice may be locally stacked high and project downward about 10–25 metres (33–80 feet) into the ocean. Besides its deterrence to the exchange of energy between the ocean and the atmosphere, the formation of sea ice generates vast quantities of cold water that help drive the circulation of the world ocean system.

Sea ice rarely forms in the open ocean below a latitude of 60° N but does occur in more southerly enclosed bays, rivers, and seas. Between about 60° and 75° N the occurrence of sea ice is seasonal, and there is usually a period of the year when the water is ice-free. Above a latitude of 75°N there is a more or less permanent ice cover. Even there, however, as much as 10 percent of the area consists of open water owing to the continual opening of leads and polynyas.

In the process of freezing, the salt in seawater is expelled as brine. The degree to which this rejection takes place increases as the rate of freezing decreases. Typically, newly formed sea ice has a salinity of 4 to 6 ‰. Even after freezing the process of purification continues but at a much slower rate. By the time the ice is one year old, it is sufficiently salt-free to be melted for drinking. This year-old, or older, salt-free sea ice is referred to as multiyear sea ice or polar pack. It can be distinguished by its smoother, rounded surface and pale blue colour. Younger ice is more jagged and grayer in colour. Because the hardness and strength of ice increases as the salts are expelled, polar pack is a special threat to shipping. First-year ice has a characteristic thickness of up to 2 metres (6 feet), whereas multiyear ice averages about 4 metres (about 12 feet) in thickness.

There is no direct evidence as to the onset of the Arctic Ocean ice cover. The origin of the ice pack was influenced by a number of factors, such as the formation of terrestrial ice caps and the interaction of the Arctic and North Atlantic waters—with their different temperature and salinity structures—with atmospheric climate variables.

What can be inferred from available data is that there was not a continuous ice cover throughout the Pleistocene Epoch (about 2.6 million to 11,700 years ago). Rather, there was a continually warm ocean until approximately 2,000,000 years ago, followed by a permanent ice pack about 850,000 years ago.

ANTARCTICA

Fifth in size among the world's continents, Antarctica's landmass is almost wholly covered by a vast ice sheet. It lies almost concentrically around the South Pole. Antarctica—the name of which means "opposite to the Arctic"—is the southernmost continent, a circumstance that has had momentous consequences for all aspects of its character.

Antarctica covers about 14.2 million square km (5.5 million square miles), and would be essentially circular except for the outflaring Antarctic Peninsula, which reaches toward the southern tip of South America (some 970 km [600 miles] away), and for two principal embayments, the Ross Sea and the Weddell Sea. These deep embayments of the southernmost Pacific and Atlantic oceans make the continent somewhat pear-shaped, dividing it into two unequal-sized parts. The larger is generally known as East Antarctica because most of it lies in east longitudes. The smaller, wholly in west longitudes, is generally called West Antarctica. East and West Antarctica are separated by the 3,060-km-long (1,900-mile-long) Transantarctic Mountains. Whereas East Antarctica consists largely of a high, ice-covered plateau, West Antarctica consists of an archipelago of mountainous islands covered and bonded together by ice.

The continental ice sheet contains approximately 29 million cubic km (7 million cubic miles) of ice, representing about 90 percent of the world's total. The average thickness is about 2.45 km (1.5 miles). Many parts of the Ross and Weddell seas are covered by ice shelves, or ice sheets floating on the sea. These shelves—the Ross Ice Shelf and the Filchner-Ronne Ice Shelf—together with other shelves around the continental margins, constitute about 10 percent of the area of Antarctic ice. Around the Antarctic coast, shelves, glaciers, and ice sheets continually calve, or discharge, icebergs into the seas.

A 2009 NASA satellite image of Antarctica, showing a vast expanse of the continent's coastline that is occupied by the Ross and Filchner-Ronne ice shelves. MODIS Rapid Response Team, NASA's Goddard Space Flight Center

Because of this vast ice, the continent supports only a primitive indigenous population of cold-adapted land plants and animals. The surrounding sea is as rich in life as the land is barren. With the decline of whaling and sealing, the only economic base in the past, Antarctica now principally exports the results of scientific investigations that lead to a better understanding of the total world environment. The present scale of scientific investigation of Antarctica began with the International Geophysical Year (IGY) in 1957–58. Although early explorations were nationalistic, leading to territorial claims, modern ones have come under the international aegis of the Antarctic Treaty. This treaty, which was an unprecedented landmark in diplomacy when it was signed in 1959 by 12 nations, preserves the continent for nonmilitary scientific pursuits.

Antarctica, the most remote and inaccessible continent, is no longer as unknown as it was at the start of IGY. All its mountain regions have been mapped and visited by geologists, geophysicists, glaciologists, and biologists. Some mapping data are now obtained by satellite rather than by observers on the surface. Many hidden ranges and peaks are known from geophysical soundings of the Antarctic ice sheets. By using radio-echo sounding instruments, systematic aerial surveys of the ice-buried terrains can be made.

The ice-choked and stormy seas around Antarctica long hindered exploration by wooden-hulled ships. No lands break the relentless force of the prevailing west winds as they race clockwise around the continent, dragging westerly ocean currents along beneath. The southernmost parts of the Atlantic, Pacific, and Indian oceans converge into a cold, oceanic water mass with singularly unique biologic and physical characteristics. Early penetration of this Southern (or Antarctic) Ocean, as it has been called, in the search for fur seals led in 1820 to the discovery of the continent. Icebreakers and aircraft now make access relatively easy, although still not without hazard in stormy conditions. Many tourists have visited Antarctica, and it seems likely that, at least in the short run, scenic resources have greater potential for economic development than do mineral and biological resources.

The term Antarctic regions refers to all areas—oceanic, island, and continental—lying in the cold Antarctic climatic zone south of the Antarctic Convergence, an important boundary with little seasonal variability, where warm subtropical waters meet and mix with cold polar waters. For legal purposes of the Antarctic Treaty, the arbitrary boundary of latitude 60° S is used. The familiar map boundaries of the continent known as Antarctica, defined as the South Polar landmass and all its nonfloating grounded ice, are subject to change with future changes of climate. The continent was ice-free during most of its lengthy geologic history, and there is no reason to believe it will not become so again in the probably distant future.

PHYSICAL FEATURES

There are two faces of the present-day continent of Antarctica. One, seen visually, consists of the exposed rock and ice-surface terrain. The other, seen only indirectly by seismic or other remote-sensing techniques, consists of the ice-buried bedrock surface. Both evolved through long and slow geologic processes.

Effects of glacial erosion and deposition dominate everywhere in Antarctica, and erosional effects of running water are relatively minor. Yet, on warm summer days, rare and short-lived streams of glacial meltwater do locally exist. The evanescent Onyx River, for example, flows from Lower Wright Glacier terminus to empty into the nondrained basin of Lake Vanda near McMurdo Sound.

Glacially sculptured landforms now predominate, as they must have some 300 million years ago, in an earlier period of continental glaciation of all of Gondwana.

With an average elevation of about 2,200 metres (7,200 feet) above sea level, Antarctica is the world's highest continent. (Asia, the next, averages about 3,000 feet.) The vast ice sheets of East Antarctica reach heights of 11,500 feet or more in four main centres: Dome A (Argus) at 81° S, 77° E; Dome C at 75° S, 125° E; Dome Fuji at 77° S, 40° E; and Vostok station at 77° S, 104° E. Without its ice, however, Antarctica would probably average little more than about 1,500 feet. It would then consist of a far smaller continent (East Antarctica) and a nearby island archipelago. A vast lowland plain between 90° E and 150° E (today's Polar and Wilkes subglacial basins) would be fringed by the ranges of the Transantarctic Mountains and of the Gamburtsev Mountains, 6,500 to 13,000 feet high. The rest might be a hilly to mountainous terrain. Relief in general would be great, with elevations ranging from 4,897 metres (16,066 feet) at Vinson Massif in the Sentinel Range, the highest point in Antarctica, to more than 8,200 feet below sea level in an adjoining marine trough to the west (Bentley Subglacial Trench). Areas that are now called "lands," including most of Ellsworth Land and Marie Byrd Land, would be beneath the sea.

Ice-scarred volcanoes, many still active, dot western Ellsworth Land, Marie Byrd Land, and sections of the coasts of the Antarctic Peninsula and Victoria Land, but principal activity is concentrated in the volcanic Scotia Arc. Only one volcano, Gaussberg (90° E), occurs along the entire coast of East Antarctica. Long dormant, Mount Erebus, on Ross Island, showed increased activity from the mid-1970s. Lava lakes have occasionally filled, but not overspilled, its crater, but the volcano's activity has been closely monitored because Antarctica's largest station

(McMurdo Station, U.S.) lies on its lower flank. One of several violent eruptions of Deception Island, a volcanic caldera, in 1967–70 destroyed nearby British and Chilean stations. Whereas volcanoes of the Antarctic Peninsula and Scotia Arc are mineralogically similar to the volcanoes typical of the Pacific Ocean rim, the others in Antarctica are chemically like those of volcanoes along the East African Rift Valley.

ANTARCTIC GLACIATION

Antarctica provides the best available picture of the probable appearance 20,000 years ago of northern North America under the great Laurentide Ice Sheet. Some scientists contend that the initial glacier that thickened over time to become the vast East Antarctic Ice Sheet originated in the Gamburtsev Mountains more than 14 million years ago. Other glaciers, such as those forming in the Sentinel Range perhaps as early as 50 million years ago, advanced down valleys to calve into the sea in West Antarctica. Fringing ice shelves were built and later became grounded as glaciation intensified. Local ice caps developed, covering West Antarctic island groups as well as the mountain ranges of East Antarctica. The ice caps eventually coalesced into great ice sheets that tied together West and East Antarctica into the single continent that is known today. Except for a possible major deglaciation as recently as 3 million years ago, the continent has been largely covered by ice since the first glaciers appeared.

Causal factors leading to the birth and development of these continental ice sheets and then to their decay and death are, nevertheless, still poorly understood. The factors are complexly interrelated. Moreover, once developed, ice sheets tend to form independent climatic patterns and thus to be self-perpetuating and eventually perhaps even self-destructing. Cold air masses draining off Antarctic lands, for example, cool and freeze surrounding oceans in winter to form an ice pack, which reduces solar energy input by increasing reflectivity and makes interior continental regions even more remote from sources of open oceanic heat and moisture. The East Antarctic Ice Sheet has grown to such great elevation and extent that little atmospheric moisture now nourishes its central part.

The volume of South Polar ice must have fluctuated greatly at times since the birth of the ice sheets. Glacial erratics and glacially striated rocks on mountain summits now high above current ice-sheet levels testify to an over-riding by ice at much higher levels. General lowering of levels caused some former glaciers flowing from the polar region through the Transantarctic Mountains to recede and nearly vanish, producing such spectacular "dry valleys" as the Wright, Taylor, and Victoria valleys near McMurdo Sound.

Doubt has been shed on the common belief that Antarctic ice has continuously persisted since its origin by the discovery reported in 1983 of Cenozoic marine diatoms—believed to date from the Pliocene Epoch (about 5.3 to 2.6 million years ago)—in glacial till of the Beardmore Glacier area. The diatoms are believed to have been scoured from young sedimentary deposits of basins in East Antarctica and incorporated into deposits of glaciers moving through the Transantarctic Mountains. If so, Antarctica may have been free or nearly free of ice as recently as about 3 million years ago, when the diatom-bearing beds were deposited in a marine sea-way. Additionally, the Antarctic Ice Sheet may have undergone deglaciations perhaps similar to those that occurred later during interglacial stages in the Northern Hemisphere. Evidence of former higher sea levels found in many areas of the Earth seems to support the hypothesis that such deglaciation occurred. If Antarctica's ice were to melt today, for example, global sea levels would probably rise about 150 to 200 feet.

The Antarctic Ice Sheet seems to be approximately in a state of equilibrium, neither increasing nor decreasing significantly according to the best estimates. Snow precipitation is offset mainly by continental ice moving seaward by three mechanisms—ice-shelf flow, ice-stream flow, and sheet flow. The greatest volume loss is by calving from shelves, particularly the Ross, Ronne, Filchner, and Amery ice shelves. Much loss also occurs by bottom melting, but this is partly compensated by a gain in mass by accretion of frozen seawater. The quantitative pattern and the balance between gain and loss are known to be different at different ice shelves, but melting probably predominates. The smaller ice shelves in the Antarctic Peninsula are currently retreating, breaking up into vast fields of icebergs, likely due to rising temperature and surface melting.

The West Antarctic Ice Sheet (WAIS) has been the subject of much research because it may be unstable. The Ross Ice Shelf is largely fed by huge ice streams descending from the WAIS along the Siple Coast. These ice streams have shown major changes—acceleration, deceleration, thickening, and thinning—in the last century or so. These alterations have affected the grounding line, where grounded glaciers lift off their beds to form ice shelves or floating glacier tongues. Changes to the grounding line may eventually transform the WAIS proper, potentially leading to removal of this ice sheet and causing a major rise in global sea level. Although the possibility of all this happening in the next 100 years is remote, major modifications in the WAIS in the 21st century are not impossible and could have worldwide effects.

These ice sheets also provide unique records of past climates from atmospheric, volcanic, and cosmic fallout; precipitation amounts and chemistry; temperatures; and even samples of past atmospheres. Thus ice-core drilling, and the subsequent analysis of these cores, has provided new information on the processes that cause climate to change. A deep coring hole at the Russian station Vostok brought up a climate and fallout history extending back more than 400,000 years. Although near the bottom, drilling has stopped because a huge freshwater lake lies between the ice and the bed at this location. Lake Vostok has probably been isolated from the atmosphere for tens of millions of years, leading to speculation of what sort of life may have evolved in this unusual setting. Research is being conducted on how to answer this question without contaminating the water body. Lake Vostok has also attracted the attention of the planetary science community, because it is a possible test site for future study of Jupiter's moon Ganymede. Ganymede possesses a layer of liquid water beneath a thick ice cover and thus has a potential for harbouring life.

Thousands of meteorites have been discovered on "blue ice" areas of the ice sheets. Only five fragments had been found by 1969, but since then more than 9,800 have been recovered, mainly by Japanese and American scientists. Most specimens appear to have landed on Antarctic ice sheets between about 700,000 and 10,000 years ago. They were carried to blue ice areas near mountains where the ancient ice ablated and meteorites became concentrated on the surface. Most meteorites are believed to be from asteroids and a few from comets, but some are now known to be of lunar origin. Other meteorites of a rare class called shergottites had a similar origin from Mars. One of these Martian shergottites has minute structures and a chemical composition that some workers have suggested is evidence for life, though this claim is very controversial.

THE SURROUNDING SEAS

The seas around Antarctica have often been likened to the moat around a fortress. The turbulent "Roaring Forties" and "Furious Fifties" lie in a circumpolar storm track and a westerly oceanic current zone commonly called the West Wind Drift, or Circumpolar Current. Warm, subtropical surface currents in the Atlantic, Pacific, and Indian oceans move southward in the western parts of these waters and then turn eastward upon meeting the Circumpolar Current. The warm water meets and partly mixes with cold Antarctic water, called the Antarctic Surface Water, to form a mass with intermediate characteristics called Subantarctic Surface Water. Mixing occurs in a shallow but broad zone of approximately 10° latitude lying south of the Subtropical Convergence (at about 40° S) and north of the Antarctic Convergence (between about 50° and 60° S). The Subtropical Convergence generally defines the northern limits of a water mass having so many unique physical and biological characteristics that it is often given a separate name, the Antarctic, or sometimes the Southern, Ocean; it contains about 10 percent of the global ocean volume.

The two convergences are well defined and important oceanic boundary zones that profoundly affect climates, marine life, bottom sedimentation, and ice-pack and iceberg drift. They are easily identified by rapid changes in temperature and salinity. Antarctic waters are less saline than tropical waters because of their lower temperatures and lesser evaporational concentration of dissolved salts. When surface waters move southward from the Subtropical Convergence zone into the subantarctic climatic belt, their temperatures drop by as much as about 9 to 16°F (5 to 9°C). Across the Antarctic Convergence, from the subantarctic into the Antarctic climatic zone, surface-water temperature drops further.

Whereas the pattern of surface currents, controlled largely by Earth's rotation, winds, water-density differences, and the geometry of basins, is relatively well understood, that of deeper water masses is more complex and less well known. North-flowing Antarctic Surface Water sinks to about 900 metres (3,000 feet) beneath warmer Subantarctic Surface Water along the Antarctic Convergence to become the Subantarctic Intermediate Water. This water mass, as well as the cold Antarctic Bottom Water, spreads far north beyond the Equator to exchange with waters of the Northern Hemisphere. The movement of the Antarctic Bottom Water is identifiable in the Atlantic as far north as the Bermuda Rise. Currents near the continent result in a circumferential belt of surface-water divergence accompanied by upwelling of deeper water masses.

Two forms of floating ice masses build out around the continent: glacier-fed semipermanent ice shelves, some of enormous size, such as the Ross Ice Shelf, and an annually frozen and melted ice pack that in winter reaches to about 56° S in the Atlantic and 64° S in the Pacific. Antarctica has been called the pulsating continent because of the annual buildup and retreat of its secondary ice-fronted coastline. Pushed by winds and currents, the ice pack is in continual motion. This movement is westward in the coastal belt of the East Wind Drift at the continent edge and eastward (farther north) at the belt of the West Wind Drift. Icebergs—calved fragments of glaciers and ice shelves—reach a northern limit at about the Subtropical Convergence. With an annual areal variation about six times as great as that for the Arctic ice pack, the Antarctic pack doubtless plays a far greater role in varying heat exchange between ocean and atmosphere and thus probably in altering global weather patterns. Long-term synoptic studies, now aided by satellite imagery, show long-period thinning in the Antarctic ice-pack regimen possibly related to global climate changes.

As part of the Deep Sea Drilling Project conducted from 1968 to 1983 by the U.S. government, the drilling ship *Glomar Challenger* undertook several cruises of Antarctic and subantarctic waters to gather and study materials on and below the ocean floor. Included expeditions were between Australia and the Ross Sea (1972–73), in the area south of New Zealand (1973), from southern Chile to the Bellingshausen Sea (1974), and two in the Drake Passage and Falkland Islands area (1974 and 1979–80). Among the ship's most significant findings were hydrocarbons discovered in sediments of Neogene and Paleogene age (some 2.6 million to 65 million years old) in the Ross Sea and rocks carried by icebergs from Antarctica found in late Oligocene sediments (those roughly 23 to 28 million years old) at numerous locations. Researchers inferred from these ice-borne debris that Antarctica was glaciated at least 25 million years ago.

Internationally funded drilling operations began in 1985 with the Ocean Drilling Program, using the new drilling vessel *JOIDES Resolution* to expand earlier *Glomar Challenger* studies. Studies in the Weddell Sea (1986–87) suggested that surface waters were warm during Late Cretaceous to early Cenozoic time and that the West Antarctic Ice Sheet did not form until about 10 million to 5 million years ago, which is much later than inferred from evidence on the continent itself. Drilling of the Kerguelen Plateau near the Amery Ice Shelf (1987–88) entailed the study of the rifting history of the Indian-Australian Plate from East Antarctica and revealed that this submerged plateau—the world's largest such feature—is of oceanic origin and not a continental fragment, as had been previously thought.

CLIMATE

The unique weather and climate of Antarctica provide the basis for its familiar appellations—Home of the Blizzard and White Desert. By far the coldest continent, Antarctica has winter temperatures that range from -128.6°F (-89.2°C), the world's lowest recorded temperature, measured at Vostok Station (Russia) on July 21, 1983, on the high inland ice sheet to -76°F (-60°C) near sea level. Temperatures vary greatly from place to place, but direct measurements in most places are generally available only for summertime. Only at fixed stations operated since the IGY have year-round measurements been made. Winter temperatures rarely reach as high as 52°F (11°C) on the northern Antarctic Peninsula, which, because of its maritime influences, is the warmest part of the continent. Mean temperatures of the coldest months are -4 to -22°F (-20 to -30°C) on the coast and -40 to -94°F (-40 to -70°C) in the interior, the coldest period on the polar plateau being usually in late August just before the return of the sun. Whereas mid-summer temperatures may reach as high as 59°F (15°C) on the Antarctic Peninsula, those elsewhere are usually much lower, ranging from a mean of about 32°F (0°C) on the coast to between -4 and -31°F (-20 and -35°C) in the interior. These temperatures are far lower than those of the Arctic, where monthly means range only from about 32°F in summer to -31°F in winter.

Wind chill—the cooling power of wind on exposed surfaces—is the major debilitating weather factor of Antarctic expeditions. Fierce winds characterize most coastal regions, particularly of East Antarctica, where cold, dense air flows down the steep slopes off interior highlands. Known as katabatic winds, they are a surface flow that may be smooth if of low velocity but that may also become greatly turbulent, sweeping high any loose snow, if a critical velocity is surpassed. This turbulent air may appear suddenly and is responsible for the brief and localized Antarctic "blizzards" during which no snow actually falls and skies above are clear. During one winter at Mirnyy Station, gusts reached more than 110 miles per hour on seven occasions. At Commonwealth Bay on the Adélie Coast the wind speed averaged 45 miles per hour (20 metres per second). Gusts estimated at between 140 and 155 miles per hour on Dec. 9, 1960, destroyed a Beaver aircraft at Mawson Station on the Mac. Robertson Land coast. Winds on the polar plateau are usually light, with monthly mean velocities at the South Pole ranging from about 9 miles per hour (4 metres per second) in December (summer) to 17 miles per hour (8 metres per second) in June and July (winter).

The Antarctic atmosphere, because of its low temperature, contains only about one-tenth of the water-vapour concentration found in temperate latitudes. This atmospheric water largely comes from ice-free regions of the southern oceans and is transported in the troposphere into Antarctica mostly in the 140° sector (80° E to 140° W) from Wilkes Land to Marie Byrd Land. Most of this water precipitates as snow along the continental margin. Rainfalls are almost unknown. Despite the tremendous volume of potential water stored as ice, Antarctica must be considered one of the world's great deserts; the average precipitation (water equivalent) is only about 2 inches (50 mm) per year over the polar plateau, though considerably more, perhaps 10 times as much, falls in the coastal belt.

Lacking a heavy and protective water-vapour-rich atmospheric layer, which in other areas absorbs and reradiates to Earth long-wave radiation, the Antarctic surface readily loses heat energy into space.

Many factors determine Antarctica's climate, but the primary one is the geometry of the Sun-Earth relationship. The 23.5° axial tilt of the Earth to its annual plane of orbit, or ecliptic, around the Sun results in long winter nights and long summer days alternating between both polar regions and causing seasonal variations in climate. On a midwinter day, about June 21, the Sun's rays reach to only 23.5° (not exact, because of refraction) from the South Pole along the latitude of 66.5° S, a line familiarly known as the Antarctic Circle. Although "night" theoretically is six months long at the geographic pole, one month of this actually is a twilight period. Only a few coastal fringes lie north of the Antarctic Circle. The amount of incoming solar radiation, and thus heat, depends additionally on the incident angle of the rays and therefore decreases inversely with latitude to reach a minimum at the geographic poles.

These and other factors are essentially the same for both polar regions. The reason for their great climatic difference primarily lies in their reverse distributions of land and sea. The Arctic is an ocean surrounded by land, while Antarctica is a continent surrounded by ocean. The Arctic Ocean, a climate-ameliorating heat source, has no counterpart at the South Pole, the great elevation and perpetually reflective snow cover of which instead intensify its polar climate. Moreover, during Antarctic winters, freezing of the surrounding sea effectively more than doubles the size of the continent and removes the oceanic heat source to nearly 3,000 km (1,800 miles) from the central polar plateau.

Outgoing terrestrial radiation greatly exceeds absorbed incoming solar radiation. This loss results in strong surface cooling, giving rise to the characteristic Antarctic temperature inversions in which temperature increases from the surface upward to about 1,000 feet above the surface. About 90 percent of the loss is replaced by atmospheric heat from lower latitudes, and the remainder by latent heat of water-vapour condensation.

Great cyclonic storms circle Antarctica in endless west-to-east procession, exchanging atmospheric heat to the continent from sources in the southern Atlantic, Pacific, and Indian oceans. Moist maritime air interacting with cold polar air makes the Antarctic Ocean in the vicinity of the Polar Front one of the world's stormiest. Few storms bring snowfalls to interior regions. With few reporting stations, weather prediction has been exceedingly difficult but is now greatly aided by satellite imagery.

Antarctica, and particularly the South Pole, attracts much interest in astronomical and astrophysical studies as well as research on the interactions between the Sun and the upper atmosphere of Earth. The South Pole is a unique astronomical location (a station from which the Sun can be viewed continuously in summer) sitting at a high geomagnetic latitude with unequaled atmospheric clarity. It possesses a thick section of pure material (ice) that can be used as a cosmic particle detector. Automatic geophysical observatories on the high polar plateau now record information on the polar ionosphere and magnetosphere, providing data that are critical to an understanding of Earth's response to solar activity.

A major focus of upper atmospheric research in Antarctica is to understand the processes leading to the annual springtime depletion in stratospheric ozone—the "ozone hole"—which has been steadily increasing since it was first detected in 1977. Ozone is destroyed as the result of chemical reactions on the surfaces of particles in polar stratospheric clouds (PSCs). These clouds are isolated within an atmospheric circulation pattern known as the "polar vortex," which develops during the long, cold Antarctic winter. The chemical reactions take place with the arrival of sunlight in spring and are facilitated by the presence of halogens (chlorine and fluorine), which are mostly products of human activity. This process of ozone destruction, which also occurs to a lesser extent in the Arctic, increases the amount of ultraviolet-B radiation reaching Earth's surface, a type of radiation shown to impair photosynthesis in plants, cause an increase in skin cancer in humans, and damage DNA molecules in living things.

THE POWER OF ICE

Even though ice covers only about 10 percent of Earth's surface, it is a significant material with a great deal of influence over life on Earth. The majority of this ice is found at or near the poles, both on top of and within the upper layers of soil and rock. During the winter, many temperate-zone lakes and rivers freeze, thawing when air temperatures rise in the spring. Ice forms from liquid water, fractures, melts, and decays. The movement of gla-cial ice over the land scours the soil from the bedrock it sits on, creating basins for future lakes, valleys for rivers, moraines, and other landforms. In the frigid seas, icebergs break away from their parent ice shelves. Some threaten shipping, and all partially modify the structure of the oceans they reside in.

As air temperatures increase from the combination of natural climate change and global warming, there is evidence of accelerated melting in mountain glaciers, permafrost regions, areas covered by sea ice, and portions of the remaining continental ice sheets. Among the numerous effects of warming temperatures, many scientists predict that sea levels will rise to flood many coastal cities and low-lying regions. Sea-level rise has happened before; it is a phenomenon tied to Earth's continuous cycling between intervals of increased cold and increased warmth.

A side from the vast continental glaciers of Antarctica and the ice floes of the Arctic Ocean, several smaller glacial features occur. Many of the smaller glaciers and ice shelves listed below are extensions of Antarctica. The only exception, the Laurentide Ice Sheet, a continental glacier that covered a large portion of northern North America during the Pleistocene Epoch (about 2.6 million to 11,700 years ago), was an Arctic phenomenon.

AMERY ICE SHELF

The Amery Ice Shelf is a large body of floating ice, in an indentation in the Indian Ocean coastline of Antarctica, west of the American Highland. It extends inland from Prydz and MacKenzie bays more than 320 km (200 miles) to where it is fed by the Lambert Glacier. The region in which the ice shelf is located was claimed by Australia in 1933. The area of this ice shelf is approximately 60,000 square km (23,200 square miles).

BEARDMORE GLACIER

This glacier in central Antarctica descends about 2,200 metres (7,200 feet) from the South Polar Plateau to Ross Ice Shelf, dividing the Transantarctic Mountains of Queen Maud and Queen Alexandra. One of the world's largest known valley glaciers, it is 200 km (125 miles) long and is 40 km (25 miles) in width. The British explorers Ernest Henry Shackleton (1908) and Robert Scott (1911) discovered the glacier on their route to the South Pole. Later scientific research found the glacier and the mountains to either side to contain petrified wood and fossils of dinosaurs, mammal-like reptiles, ferns, and coral—evidence of a time when Antarctica possessed a temperate climate.

FILCHNER ICE SHELF

This large body of floating ice lies at the head of the Weddell Sea, which is itself an indentation in the Atlantic coastline of Antarctica. It is more than 200 metres (650 feet) thick and has an area of 260,000 square km (100,400 square miles). The shelf extends inland on the east side of Berkner Island for more than 400 km (250 miles) to the escarpment of the Pensacola Mountains. The name Filchner was originally applied to the whole shelf, including the larger area west of Berkner Island now called the Ronne Ice Shelf. Because of this, and the fact that the two shelves can be separated only at Berkner Island, the name Filchner-Ronne Ice Shelf is frequently applied to the whole ice mass. The ice shelf, named for the German explorer Wilhelm Filchner, was claimed by the United Kingdom (1908) and by Argentina (1942). Argentina, the United Kingdom, and the United States have operated research stations along its northern edge.

LARSEN ICE SHELF

The Larsen A Ice Shelf disintegrated in 1995, whereas the Larsen B Ice Shelf broke apart in 2002. Both events were caused by water from surface melting that ran down into crevasses, refroze, and wedged each shelf into pieces.

This ice shelf in the northwestern Weddell Sea adjoins the east coast of the Antarctic Peninsula and is named for Captain Carl A. Larsen, who sailed along the ice front in 1893. It originally covered an area of 86,000 square km (33,000 square miles), excluding the numerous small islands within the ice shelf. The shelf was narrow in its southern half but gradually widened toward the Antarctic Circle to the north before narrowing again. As air temperatures over the Antarctic Peninsula warmed slightly in the second half of the 20th century, the Larsen shelf shrank dramatically. In January 1995 the northern portion (known as Larsen A) disintegrated, and a giant iceberg calved from the middle section (Larsen B). Larsen B steadily retreated until February–March 2002, when it too collapsed and disintegrated. These events left the Larsen Ice Shelf covering only 40 percent of its former area.

LAURENTIDE ICE SHEET

The principal glacial cover of North America during the Pleistocene Epoch (about 2.6 million to 11,700 years ago), it spread as far south as latitude 37° N and covered an area of more than 13,000,000 square km (5,000,000 square miles) at its maximum extent. In some areas its thickness reached 2,400–3,000 metres (8,000–10,000 feet) or more. The Laurentide Ice Sheet probably originated on the Labrador-Ungava plateau and on the mountains of the Arctic islands of Canada, and centred over Hudson Bay. As it spread, the glacial ice mass appears to have combined with other ice caps that had formed on local highlands in eastern Canada and in the northeastern United States.

RONNE ICE SHELF

The Ronne Ice Shelf is a large body of floating ice, lying at the head of the Weddell Sea, which is itself an indentation in the Atlantic coastline of Antarctica. More than 150 metres (500 feet) thick and extending inland for more than 840 km (520 miles), it lies immediately west of Filchner Ice Shelf, from which it is partially separated by Berkner Island. Often the names of the two ice shelves are combined as the Filchner-Ronne Ice Shelf. This name is appropriate, because they are only partly separated by Berkner Island, and the combined ice shelf originally bore the name Filchner. The Filchner-Ronne Ice Shelf has a combined area of about 422,000 square km (163,000 square miles), making it one of the two largest ice shelves on Earth. Radio-echo sounding and coring suggest that a substantial amount of the ice thickness was formed by the accretion of ice crystals from cooled seawater below. Oceanographic data, however, suggest an average net loss by subshelf melting of 290–300 mm (11.4–11.8 inches) per year. Named for Edith Ronne, wife of the American explorer Finn Ronne, the ice shelf was claimed by the United Kingdom (1908), Chile (1940), and Argentina (1942).

SHACKLETON ICE SHELF

This sheet of floating ice borders the Queen Mary Coast, Antarctica, on the Indian Ocean. It was discovered and named for Ernest Shackleton, the British explorer, by Douglas Mawson's expedition, 1911–14. It lies between the main Russian Antarctic station Mirnyy and the Polish station Dobrowlowski.

SKELTON GLACIER

This Antarctic glacier is situated on the Hillary Coast of Victoria Land, to the northeast of the Cook Mountains, near McMurdo Sound. It flows sluggishly southward into the Ross Ice Shelf. The greatest known thickness of ice along its 62-km (39-mile) length occurs at a point about 48 km (30 mi) from the tip of its floating terminus. There the ice is about 1,450 metres (4,760 feet) thick. The east side of the glacier is covered by a discontinuous belt of graywacke-like metasedimentary and metavolcanic rock.

WILKINS ICE SHELF

The Wilkins Ice Shelf is a large body of floating ice covering the greater part of Wilkins Sound off the western coast of the Antarctic Peninsula. Both the ice shelf and the sound were named for Australian-born British explorer Sir George Hubert Wilkins, who first scouted the region by airplane in late December 1928. The Wilkins Ice Shelf spanned the region between Alexander Island, Charcot Island, and Latady Island in the Bellingshausen Sea, an area of about 16,000 square km (6,200 square miles), before its retreat began in the late 1990s. By the early 21st century the ice shelf had substantially diminished because of rising regional air temperatures and the physical stresses of ocean wave activity. In January 2008 the ice shelf covered an area of approximately 13,700 square km (about 5,300 square miles). However, a section measuring 405 square km (about 160 square miles) collapsed by March of that year, leaving a thin bridge of continuous ice connecting the ice shelf to Charcot Island. This bridge, only about 6 km (3.7 miles) wide at its widest point, acted like a dam to hold back the shelf's partially broken interior from the open sea. In April 2009 the ice bridge lost its connection to Charcot Island, increasing the likelihood of rapid disintegration of the remaining ice shelf.

albedo The fraction of light reflected by a body or surface.

anchor ice Ice formed beneath the surface of water and attached to the bottom of a water body or to submerged objects.

calving The separation of a volume of ice from its parent glacier.

candling Another name for ice deterioration, so-called because of the similarity of deteriorating ice crystals to an assembly of closely packed candles.

cirque An amphitheatre-shaped basin characterized by steep walls, at the head of a glacial valley.

crevasse A fissure or crack in a glacier produced by stress associated with the movement of a glacier.

drumlin An oval or elongated hill thought to result from the streamlined movement of glacial ice sheets across rock debris, or till.

esker A long, narrow, winding ridge composed of stratified sand and gravel deposited by a subglacial or englacial meltwater stream.

felsenmeer Exposed rock surfaces that have been quickly broken up by frost action so that much rock is buried under a cover of angular shattered boulders.

fjord A long narrow arm of the sea, commonly extending far inland, that results from marine inundation of a glaciated valley.

frazil Ice that forms as small plates drifting in rapidly flowing water where it is too turbulent for pack ice to form.

gelifluction The process by which the active layer of permafrost moves under the influence of gravity.

hummock A rounded ridge of ice.

isothermal Relating to or marked by changes of volume or pressure under conditions of constant temperature.

lee The side that is sheltered from the wind.

moraine An accumulation of rock debris (till) carried or deposited by a glacier.

nucleation The process by which a small number of ions, atoms, or molecules become arranged in a pattern characteristic of a crystalline solid, forming a site upon which additional particles are deposited as the crystal grows.

ogive A graph of a cumulative distribution function or a cumulative frequency distribution.

percolation zone The area on a glacier or ice sheet characterized by limited surface melting occurs, but the meltwater refreezes in the same snow layer.

pingo A dome-shaped hill that rises in permafrost regions as a result of the hydrostatic pressure of freezing ground-water.

plucking Glacial action that results in the removal of larger pieces of rock from the glacier bed; also known as quarrying.

polynyas Semipermanent areas of open water in sea ice.

relict A relief feature or rock that remains after other parts have disappeared.

roches moutonnées Bedrock knobs or hills that have a gently inclined, glacially abraded, and streamlined stoss side and a steep, glacially plucked lee side.

sastrugi Jagged erosional features (often cut into snow dunes) caused by strong prevailing winds that occur after snowfall.

shergottites Igneous volcanic rock meteorites.

shoal An accumulation of sediment in a river channel or on a continental shelf that is potentially dangerous to ships.

solifluction The flowage of water-saturated soil down a steep slope.

stoss Facing toward the direction from which an over-riding glacier impinges.

supercooling When water remains liquid below it's normal freezing point of 0 °C (32 °F).

talus A slope formed especially by an accumulation of rock debris.

thermokarst Land-surface configuration that results from the melting of ground ice in a region underlain by permafrost.

till Material laid down directly or reworked by a glacier.

ICE

Samuel C. Colbeck (ed.), *Dynamics of Snow and Ice Masses* (1980), includes chapters on valley glaciers, ice sheets, snow packs, icebergs, sea ice, and avalanches, emphasizing the basic physics. P.V. Hobbs, *Ice Physics* (1974); and Victor F. Petrenko and Robert W. Whitworth, *Physics of Ice* (1999), treat all aspects of the physics and chemistry of ice. W. Richard Peltier (ed.), *Ice in the Climate System* (1993), is a modern review of the past, present, and future interactions between ice and climate.

PERMAFROST

A.L. Washburn, *Geocryology* (1979), is the most thorough book in English on permafrost and periglacial processes. H.M. French, *The Periglacial Environment* (1976), clearly summarizes permafrost and periglacial processes, with emphasis on examples from Canada. The greatest source of permafrost information is the proceedings of the various International Conference on Permafrost meetings; each volume contains numerous up-to-date papers in English from many different countries. Wilfried Haeberli, *Creep of Mountain Permafrost: Internal Structure and Flow of Alpine Rock Glaciers* (1985), is a discussion of rock glaciers, a prominent feature of Alpine permafrost. Troy L. Péwé, "Alpine Permafrost in the Contiguous United States: A Review," *Arctic and Alpine Research,* 15(2):145–156 (1983), summarizes in detail the character and distribution of mountain permafrost in this region.

Arthur H. Lachenbruch, *Mechanics of Thermal Contraction Cracks and Ice-Wedge Polygons in Permafrost* (1962), is a classic paper on the quantitative interpretation of the formation of ice-wedge polygons in permafrost. Troy L. Péwé, Richard E. Church, and Marvin J. Andresen, *Origin and Paleoclimatic Significance of Large-Scale Patterned Ground in the Donnelly Dome Area, Alaska* (1969), discusses the origin of ice-wedge casts and relict permafrost in central Alaska and offers paleoclimatic interpretations. R. Dale Guthrie, *Frozen Fauna of the Mammoth Steppe* (1990), discusses fossil carcasses of Ice Age mammals preserved in permafrost. Troy L. Péwé, *Geologic Hazards of the Fairbanks Area, Alaska* (1982), a highly illustrated work, contains an up-to-date presentation of the greatest geologic hazard to life in polar areas: problems posed by seasonally and perennially frozen ground. G.H. Johnston (ed.), *Permafrost: Engineering Design and Construction* (1981), is a comprehensive book on construction problems in permafrost areas, with examples mainly from northern Canada. Troy L. Péwé, "Permafrost," in George A. Kiersch et al. (eds.), *The Heritage of Engineering Geology: The First Hundred Years* (1991), pp. 277–298, provides an up-to-date, well-illustrated treatment of the origin, distribution, and ice content of permafrost and of engineering problems in permafrost regions.

ICE IN LAKES AND RIVERS

E.R. Pounder, *The Physics of Ice* (1965), treats concisely the structure and physical properties of ice. George D. Ashton (ed.), *River and Lake Ice Engineering* (1986), provides a comprehensive treatment of the general principles for engineering applied to river and lake ice problems. Bernard Michel, *Winter Regime of Rivers and Lakes* (1971), treats freshwater ice. The essay by George D. Ashton, "Freshwater Ice Growth and Decay," in Samuel C. Colbeck (ed.), *Dynamics of Snow and Ice Masses* (1980), summarizes the general principles of river ice behaviour, including ice accumulation processes and thermal effects. For information regarding the geographic distribution of lake and river ice, the following works are useful: W.T.R. Allen, *Freeze-up, Break-up, and Ice Thickness in Canada* (1977?); and, for the former Soviet Union, I.P. Gerasimov et al. (eds.), *Fiziko-geograficheskii Atlas Mira* (1964); and L.N. Mesiatseva, *Atlas SSSR* (1984).

GLACIERS

A beautifully illustrated introduction to glaciers is contained in Austin Post and E.R. LaChapelle, *Glacier Ice*, rev. ed. (2000). W.S.B. Paterson, *The Physics of Glaciers*, 3rd ed. (1994), is the standard text on glaciers and ice sheets, emphasizing process rather than description. J.T. Andrews, *Glacial Systems* (1975), is a compact, clear introduction to glaciers and their environment. Michael Hambrey and Jürg Alean, *Glaciers* (1992); and Robert P. Sharp, *Living Ice* (1988), are well-illustrated introductions to glaciers and their effect on the landscape. National Research Council (U.S.), Ad Hoc Committee on the Relationship between Land Ice and Sea Level,
   Glaciers, Ice Sheets, and Sea Level: Effect of a CO_2-Induced Climatic Change (1985); and D.J. Drewry (ed.), *Antarctica: Glaciological and Geophysical Folio* (1983), is a large-format compendium on the Earth's largest ice mass. "Fast Glacier Flow: Ice Streams, Surging, and Tidewater Glaciers," *Journal of Geophysical Research*, part B, *Solid Earth and Planets*, 92(9):8835–8841 (1987), is a collection of review

papers and scientific contributions from the Chapman
Conference on Fast Glacier Flow.

GLACIAL LANDFORMS

The classic text on glacial geology is Richard Foster Flint,
Glacial and Quaternary Geology (1971), encyclopaedic coverage including an extensive bibliography. Recent hypotheses
and observations on glacial erosion and deposition are
included in David Drewry, *Glacial Geologic Processes* (1986),
even though the coverage of glacial landforms is not complete. David E. Sugden and Brian S. John, *Glaciers and
Landscape: A Geomorphological Approach* (1976, reprinted
1984), is an excellent detailed introduction to glacial landforms and the processes that shaped them. More
theoretical emphasis can be found in Clifford Embleton
and Cuchlaine A.M. King, *Glacial Geomorphology*, 2nd ed.
(1975), and *Periglacial Geomorphology*, 2nd ed. (1975). A.L.
Washburn, *Geocryology: A Survey of Periglacial Processes and
Environments* (1979), contains numerous explanatory pho-
tographs and diagrams. A collection of articles is found in
Cuchlaine A.M. King (ed.), *Periglacial Processes* (1976). The
most comprehensive and up-to-date account of glacial
geomorphology and sedimentology is Douglas I. Benn
and David J.A. Evans, *Glaciers and Glaciation* (1998). A
detailed discussion of the formation of permafrost can be
found in Stuart A. Harris, *The Permafrost Environment*
(1986), and in Peter J. Williams and Michael W. Smith, *The
Frozen Earth: Fundamentals of Geocryology* (1989).

ICEBERGS AND SEA ICE

Michael Hambrey and Jürg Alean, *Glaciers*, 2nd ed. (2004),
provides extensive background on glaciers and the processes that give birth to icebergs. Peter Wadhams, *Ice in the*

Ocean (2000), offers a moderately technical treatment of icebergs, sea ice, and the impacts of both on Earth's climate system. Methods of detecting icebergs by surface, airborne, and satellite radar are described in Simon Haykin et al. (eds.), *Remote Sensing of Sea Ice and Icebergs* (1994). A comprehensive resource that charts the distribution of icebergs emitted from islands in the Russian Arctic is Valentin Abramov, *Atlas of Arctic Icebergs: The Greenland, Barents, Kara, Laptev, East-Siberian, and Chukchi Seas and the Arctic Basin*, ed. by Alfred Tunik (1996). Other literature on the physical and climatic properties of icebergs is found in specialist journals such as *Journal of Geophysical Research* (weekly); *The Journal of Glaciology* (quarterly); and *Cold Regions Science and Technology* (quarterly).

SEA ICE

David N. Thomas, *Frozen Oceans: The Floating World of Pack Ice* (2004), offers an account designed to appeal to the more general, scientifically literate reader. A moderately technical but wide-ranging treatment of sea ice based on the author's extensive fieldwork in cold regions is Peter Wadhams, *Ice in the Ocean* (2000, reprinted 2002). A more rigorous, academic perspective of sea ice is found in David N. Thomas and Gerhard S. Dieckmann (eds.), *Sea Ice: An Introduction to Its Physics, Chemistry, Biology, and Geology* (2003). I.A. Melnikov, *The Arctic Sea Ice Ecosystem* (1997), describes the structure, composition, and dynamics of the Arctic sea ice ecosystem from historical observations and measurements made at Soviet North Pole stations. Comprehensive portrayals of the process of active and passive microwave remote sensing of sea ice may be found in Frank D. Carsey (ed.), *Microwave Remote Sensing of Sea Ice* (1992); and Dan Lubin and Robert Massom, *Polar Remote Sensing, Volume 1: Atmosphere and Oceans* (2005).

THE ARCTIC

GENERAL WORKS

Appealing, broad, and beautifully illustrated surveys of the region are offered in Fred Bruemmer et al., *The Arctic World* (1985), a descriptive work; and Steven B. Young, *To the Arctic: An Introduction to the Far Northern World* (1989), a more scholarly guide focusing on natural history. State-of-the-art writing on all aspects of Arctic research, of various levels of sophistication, is found in such periodicals as *Arctic* (quarterly), a journal of the Arctic Institute of North America; *Arctic Science, Engineering, and Education Awards* (annual), published by the National Science Foundation of the United States; and *Polar Record* (quarterly). The region is also considered in special thematic issues of various publications, as in "The Arctic Ocean," a separate issue of the quarterly *Oceanus*, vol. 29, no.1 (Spring 1986).

THE LAND

David Sugden, *Arctic and Antarctic: A Modern Geographical Synthesis* (1982), offers a general introduction to the geography of the polar lands. Axel Somme (ed.), *A Geography of Norden: Denmark, Finland, Iceland, Norway, Sweden*, new ed. (1968), is a broad regional geography; and a particular region is studied in Clyde Wahrhaftig, *Physiographic Divisions of Alaska* (1965); and Howel Williams (ed.), *Landscapes of Alaska: Their Geologic Evolution* (1958).

R.J. Fulton (ed.), *Quaternary Geology of Canada and Greenland* (1989), discusses relief geology, with relevant chapters discussing the ice-free areas and inland ice of Greenland and the characteristics of the Queen Elizabeth Islands. William L. Graf (ed.), *Geomorphic Systems of North America* (1987), is a study of the landforms of northern countries, including the Arctic Lowlands of Canada and Alaska; and A.A. Velichko, H.E. Wright, Jr., and C.W. Barnosky (eds.), *Late Quaternary Environments of the Soviet Union*, trans. from Russian (1984), features relevant regional information scattered in several chapters. Soils, drainage, and glaciation are discussed in J. Ross Mackay, "The World of Underground Ice," *Annals of the Association of American Geographers* 62(1):1–22 (March 1972); Peter J. Williams and Michael W. Smith, *The Frozen Earth* (1989); H.M. French, *The Periglacial Environment* (1976); and Jack D. Ives and Roger G. Barry (eds.), *Arctic and Alpine Environments* (1974), examining physical environments, the biota, humans' reaction to and impact on them, and the northern climates. Climatic data on the Arctic are found in S. Orvig (ed.), *Climates of the Polar Regions* (1970); and Jean M. Grove, *The Little Ice Age* (1988).

Barry Lopez, *Arctic Dreams: Imagination and Desire in a Northern Landscape* (1986), is an accurate though imaginative survey of natural history. Bryan Sage et al., *The Arctic & Its Wildlife* (1986), presents a well-illustrated survey of the environment. Nicholas Polunin, *Introduction to Plant Geography and Some Related Sciences* (1960), includes a chapter on plant life in cold climates. Other texts that discuss the Arctic biome include M.J. Dunbar, *Ecological Development in Polar Regions: A Study in Evolution* (1968); Bernard Stonehouse, *Animals of the Arctic: The Ecology of the Far North* (1971); Fred Bruemmer, *Arctic Animals: A Celebration of Survival* (1987); and Rita A. Horner (ed.), *Sea Ice Biota* (1985).

THE ARCTIC OCEAN

Appealing, broad, and beautifully illustrated surveys of the region are offered in Fred Bruemmer et al., *The Arctic World* (1985), a descriptive work; and Steven B. Young, *To the Arctic: An Introduction to the Far Northern World* (1989), a more scholarly guide focusing on natural history. State-of-the-art writing on all aspects of arctic research, of various levels of sophistication, is found in such periodicals as *Arctic* (quarterly), a journal of the Arctic Institute of North America; *Arctic Science, Engineering, and Education Awards* (annual), published by the National Science Foundation of the United States; and *Polar Record* (quarterly). See also "The Arctic Ocean," a separate issue of the quarterly *Oceanus*, vol. 29, no.1 (Spring 1986).

For geology and geophysics of the region, see Arthur Grantz, L. Johnson, and J.F. Sweeney (eds.), *The Arctic Ocean Region* (1990), which examines mainly the geology of the ocean floor and adjoining lands, with discussions of physical environments; Yvonne Herman (ed.), *Marine Geology and Oceanography of the Arctic Seas* (1974), and *The Arctic Seas: Climatology, Oceanography, Geology, and Biology* (1989); Burton G. Hurdle (ed.), *The Nordic Seas* (1986); and H.R. Jackson and G.L. Johnson, "Summary of Arctic Geophysics," *Journal of Geodynamics* 6(1–4):245–262 (1986).

Works on oceanography and ice behaviour include K. Aagaard, "On the Deep Circulation in the Arctic Ocean," *Deep-Sea Research*, part A, 28(3A):251–268 (March 1981); K. Aagaard and E.C. Carmack, "The Role of Sea Ice and Other Fresh Water in the Arctic Circulation," *Journal of Geophysical Research* 94(C10):14485–14498 (October 1989); *Arctic Research Advances and Prospects: Proceedings of the Conference of Arctic and Nordic Countries on Coordination of Research in the Arctic*, 2 vol. (1990); L.K. Coachman, K. Aagaard, and R.B. Tripp, *Bering Strait: The Regional Physical Oceanography* (1975); W.D. Hibler, III, and K. Bryan, "A Diagnostic Ice-Ocean Model," *Journal of Physical Oceanography* 17(7):987–1015 (July 1987); *Sea-Ice and Climate: Report of the Fourth Session of the Working Group on Sea-Ice and Climate* (1990), a special publication of the World Meteorological Organization and the International Council of Scientific Unions; G.A. Maykut, "Large-Scale Heat Exchange and Ice Production in the Central Arctic," *Journal of Geophysical Research* 87(C10):7971–7984 (September 1982); and Norbert Untersteiner, *The Geophysics of Sea Ice* (1986). Louis Rey (ed.), *The Arctic Ocean: The Hydrographic Environment and the Fate of Pollutants* (1982), collects papers on the marine pollution in the region.

ANTARCTICA

GENERAL WORKS

John Stewart, *Antarctica: An Encyclopedia*, 2 vol. (1990), emphasizes history and geography but includes entries on geologic features and scientific topics, as well as a lengthy, annotated bibliography. The most complete guide to literature about the Antarctic is the U.S. Library of Congress, *Antarctic Bibliography* (annual). Fred G. Alberts (compiler and ed.), *Geographic Names of the Antarctic* (1981), contains a compilation and derivation of Antarctica's place names up to 1979, with coordinates, details of discovery, and for whom each was named. E.I. Tolstikov, *Atlas Antarktiki*, 2 vol. (1966–69), is a comprehensive map collection in Russian, useful especially when complemented by the translation of legend matter and explanatory text from vol. 1, published as "Atlas of Antarctica," a special issue of *Soviet Geography: Review & Translation*, vol. 8, no. 5–6 (May–June 1967).

Louis O. Quam (ed.), *Research in the Antarctic* (1971); Richard S. Lewis and Philip M. Smith (eds.), *Frozen Future: A Prophetic Report from Antarctica* (1973); and D.W.H. Walton (ed.), *Antarctic Science* (1987), contain review articles by leading experts on most subjects of research, the latter two also with articles on resources, economics, politics, and the outlook for the future. Richard Fifield, *International Research in the Antarctic* (1987), introduces the various types of research undertaken in Antarctica. A general review, Raymond Priestley, Raymond J. Adie, and G. De Q. Robin (eds.), *Antarctic Research* (1964), emphasizes British scientific achievements, particularly in the Antarctic Peninsula and Scotia Arc. Semitechnical to nontechnical reviews of current projects and exploration are in summary articles in the *Antarctic Journal of the United States* (quarterly). A general nontechnical review of earlier research is provided in the still-useful work by H.G.R. King, *The Antarctic* (1969). G.E. Fogg, *A History of Antarctic Science* (1992), traces the development of scientific inquiry in Antarctica. *Antarctic Science* (quarterly) covers all fields of scientific research on the continent. American Geographical Society of New York, *Antarctic Map Folio Series*, 19 vol. (1964–75); and American Geophysical Union, *Antarctic Research Series* (irregular), provide modern maps and technical accounts of all phases of the research programs. A more recent compilation of data on Antarctica, displayed as maps and tables, is D.J. Drewry (ed.), *Antarctica: Glaciological and Geophysical Folio* (1983).

F.M. Auburn, *Antarctic Law and Politics* (1982), provides a comprehensive discussion of the legal aspects of the Antarctic Treaty, jurisdictional problems of crime, ecology, and tourism. Gillian D. Triggs (ed.), *The Antarctic Treaty Regime: Law, Environment, and Resources* (1987), discusses current aspects of issues raised by the Antarctic Treaty. Anthony Parsons, *Antarctica: The Next Decade* (1987), addresses the history of the Antarctic Treaty and its future as well as current and projected uses of the continental region.

PHYSICAL GEOGRAPHY

W.N. Bonner and D.W.H. Walton (eds.), *Antarctica* (1985); and R.M. Laws (ed.), *Antarctic Ecology*, 2 vol. (1984), discusses the continent's physical environment, fauna, flora, land and sea ecology, conservation, and exploitation. I.B. Campbell and G.G.C. Claridge, *Antarctica: Soils, Weathering Processes, and Environment* (1987), provides a summary of recent research.

GLACIERS AND SEAS

Malcolm Mellor (ed.), *Antarctic Snow and Ice Studies* (1964); and A.P. Crary (ed.), *Antarctic Snow and Ice Studies II* (1971), are collections mainly of topical studies of greatly varied scope. John Mercer, *Glaciers of the Antarctic* (1967), provides a general review of Antarctica's glaciers. Stephen J. Pyne, *The Ice* (1986), investigates Antarctic ice and also describes other aspects of the region, including exploration, literature, and art. A.L. Gordon and R.D. Goldberg, *Circumpolar Characteristics of Antarctic Waters* (1970); and Joseph L. Reid and Dennis E. Hayes (eds.), *Antarctic Oceanology*, 2 vol. (1971–72), describe features of South Polar water masses, their currents, and their interactions with subtropical and subantarctic waters, as well as of the ocean floor and sediment carpet. George Deacon, *The Antarctic Circumpolar Ocean* (1984), includes a summary of early discoveries by explorers, sealers, and whalers and a review of modern knowledge of Antarctic waters. Other works include Martin Jeffries (ed.), *Antarctic Sea Ice: Physical Processes, Interactions, and Variability* (1998); and Stanley S. Jacobs and Raymond F. Weiss (eds.), *Ocean, Ice, and Atmosphere: Interactions at the Antarctic Continental Margin* (1998).

INDEX